Remarkable Agaves
and Cacti

REMARKABLE
AGAVES
AND CACTI

PARK S. NOBEL
University of California,
Los Angeles

New York   Oxford
OXFORD UNIVERSITY PRESS
1994

Oxford University Press

Oxford   New York   Toronto
Delhi   Bombay   Calcutta   Madras   Karachi
Kuala Lumpur   Singapore   Hong Kong   Tokyo
Nairobi   Dar es Salaam   Cape Town
Melbourne   Auckland   Madrid

and associated companies in
Berlin   Ibadan

Published by Oxford University Press, Inc.,
200 Madison Avenue, New York, New York 10016

Library of Congress Cataloging-in-Publication Data
Nobel, Park S.
Remarkable agaves and cacti / Park S. Nobel.
p.   cm.   Includes bibliographical references and index.
ISBN 0-19-508414-4
ISBN 0-19-508415-2 (pbk).
1. Agave.   2. Cactus.   3. Agave—Utilization.
4. Cactus—Utilization.
I. Title
QK495.A26N64   1994   584'.43—dc20   93-20756

2 4 6 8 9 7 5 3 1
Printed in the United States of America
on acid-free paper

Preface

What comes to mind when you think of an agave or a cactus? Perhaps you picture the magnificent saguaro (*Carnegiea gigantea*), the state flower of Arizona and the enduring symbol of the Sonoran Desert. Maybe you envision the tall showy flowering stalk of a century plant (*Agave americana*) in a botanical garden. Or perhaps you imagine *Mammillaria lasiacantha* or other diminutive cacti tolerating high temperatures and extremely long droughts in the Chihuahuan Desert of Mexico. But how about the little prickly pear (*Opuntia fragilis*), which is capable of surviving extremely low temperatures of $-40°C$ ($-40°F$) in western Canada? Or consider the large fields of sisal (*Agave sisalana*), cultivated for fiber in eastern Africa, and the prickly pear cactus (*Opuntia ficus-indica*), cultivated for fruit and cattle fodder in many countries. Despite their reputation as slow growers that tolerate drought and neglect, various species of agaves and cacti have recently been shown to be extremely productive.

This book examines three aspects of agaves and cacti. First, their past uses and present-day commercial values are considered. Evidence based on fossilized human feces indicates that agaves and cacti have been part of the human diet for over 9000 years. In the nineteenth century, an incredibly valuable red dye was obtained from an insect feeding on a prickly pear cactus in the Canary Islands, and today many beverages are produced from agaves in Mexico and the fruits of many species of cacti are sold in markets worldwide. Second, their physiology—the science of how they function—is key to understanding agaves and cacti. Physiology is considered at the cellular level—Why are agaves and cacti so adept at conserving water?—and at the organ level—What are the unique characteristics of the roots and shoots of these desert succulents? Third, the production of plant biomass, which is a consequence of the physiological activity of the roots and shoots, can be higher for certain cultivated agaves and cacti than for nearly all other crops. The basis for the high possible productivity can be explained at a cellular level. This knowledge, coupled with a rediscovery of the traditional uses of agaves and cacti, could revolutionize their utilization in the twenty-first century.

Scientists are becoming increasingly specialized and tend to communicate their findings primarily with their peers. Yet scientific discoveries can be of interest to a wide audience and can profoundly affect economic and political decisions, especially if jargon and burdensome technical detail are avoided. A great many people in all parts of the world, amateurs and professionals alike, are already intrigued by succulent plants. I hope to stimulate even more interest in agaves and cacti and to present scientific information pertinent to understanding the unique physiology of these remarkable plants.

The book is designed for a diverse audience. Those readers seeking a cultural appreciation of agaves and cacti may focus on Chapters 1 through 3 and 8. Those interested in the physiology of these plants can concentrate on Chapters 4 through 7. And readers in a hurry may begin with Chapter 8, looking up certain details using the many figures, tables, and plates that are cross-referenced in that chapter. I hope, however, that many will enjoy the whole book in its intended order, as terms and concepts are introduced gradually to create a greater appreciation of many facets of agaves and cacti. For convenience of reference, the Index indicates those pages in boldface on which a term or an object is defined or illustrated.

Financial support for this book from the Environmental Sciences Division, Office of Health and Environmental Research, United States Department of Energy, is gratefully acknowledged. Many helpful criticisms were provided by Judy Amos, Edmundo García-Moya, Eric Graham, Gabriella Gray, Michael Loik, Patsy Miller, and Gretchen North. Judy Amos did an excellent typing job; Eric Graham helped prepare the Index; Margaret Kowalczyk prepared the drawings; and Gretchen North helped prepare the photographs.

Los Angeles   P.S.N.
November 30, 1993

Contents

Color plates follow page 86

1. **Introduction, 3**
   Classification and Evolution, 4
   Distribution: Threatened Species, 7
   Features in Native Environments, 10
      Pollination, 10
      Vegetative Propagation, 12
      Age and Propagation of Agave deserti, *14*
      Tilting of Barrel Cacti, 14
   Human Uses, 16
   Morphology and Anatomy, 20
   Physiology: Water Conservation, 24
   References, 27

2. **Agaves: Food, Beverages, and Fiber, 28**
   Food, 29
   Aguamiel and Pulque, 32
   Mescal and Tequila, 34
   Fiber, 38
      Mexico, 38
      Eastern Africa, 40

Other Uses, 41
  Ornamental Horticulture, 41
  Sapogenins, 42
References, 44

3. Cacti: Many Uses . . . and a Few Mistakes, 45

Fruits, 46
  Cactus Pears, 47
  The Sicilian Experience, 49
  Other Species, 51
Young Cladodes as Vegetables, 53
Forage and Fodder, 55
Royal Red: The Cochineal Story, 58
Peyote and Other Hallucinogens, 60
Other Uses, 61
  Candy, 63
  Ornamentals, 63
  Medicine and Hormones, 64
When Cacti Take Over, 65
  Australia, 65
  South Africa, 67
  A North American Experience, 68
References, 71

4. Roots: Water Uptake, 73

Root Distribution, Morphology, and Anatomy, 74
Water Potential: Hydraulic Conductivity, 77
Water Movement, 80
Other Root Properties, 82
  Root:Shoot Ratios, 83
  Mycorrhizae, 84
  Nutrients, 85
  Salinity, 86
References, 87

5. Shoots: Environmental Interactions, 89

Water Relations, 89
  Water Storage, 90

The Epidermis, 91
Drought Tolerance, 93
Radiation, 94
Light Distribution over Agaves, 94
Cactus Morphology, 95
Cladode Orientation, 95
Temperature Relations, 97
Morphology, 98
Low Temperature, 100
High Temperature, 103
Nurse Plants, 104
References, 106

6. CO_2 Uptake by Plants, 108
Binding of CO_2, 109
Three Different Pathways, 110
Cellular Location, 112
Energetics of CO_2 Fixation, 114
Cellular Energy Currencies, 115
Photorespiration, 116
Energetic Costs, 118
Daily Patterns of CO_2 Exchange, 120
Water-Use Efficiency, 124
References, 126

7. Plant Productivity, 128
Environmental Influences on CO_2 Uptake, 129
Water, 129
Temperature, 130
Photosynthetic Photon Flux, 132
An Environmental Productivity Index, 134
The Basic Equation, 134
Elevational Effects, 135
Importance of Plant Spacing, 137
Highest Productivities, 138
CAM Plants, 139
C_3 and C_4 Plants, 141
Interactions Among Organs, 143
References, 145

8. The Future, 147

Conservation, 148
Agaves, 151
Cacti, 152
Global Climate Change, 156
Predictions, 157

Index, 161
Index, 161

Remarkable Agaves
and Cacti

1

Introduction

Remarkable—meaning worthy of notice, uncommon, even extraordinary—aptly describes agaves and cacti. They survive in all kinds of hostile environments, from hot dry deserts to snowy mountains. Agaves and cacti have been part of the human diet for at least 9000 years. Are changes now threatening their place in nature? Human influences can range from the harvest of some fruits and the collection of prized plants to large-scale habitat destruction. Indeed, the value of plants to society often dictates their survival in the natural world. The vast ethnobotanical history—the study of plants as they relate to people—of agaves and cacti spans the collecting of a liquid from an agave for fermentation (Figure 1.1) to the legendary perching of an eagle on a cactus, designating the center of the mighty Aztec empire and now depicted on the flag of Mexico (Plate A). The Aztec legend led to a banner called *nopantli*, leading to the term *nopal,* which is now commonly applied to the type of cactus on which the eagle perched.

Various species of cacti are collected, bred, and then sold around the world as ornamental houseplants that require minimal care and can produce showy flowers. Other species of cacti are raised in 25 countries for their fruit, and several species of agaves are used to produce beverages. In essentially all cases, these plants can tolerate long periods without water. Tolerance of drought by agaves and cacti is influenced by their *morphology* (the external form), their *anatomy* (the internal form, including sizes and shapes of cells), and their *physiology*. The word *physiology* is derived from the Greek *physi*, meaning nature, and −*logia*, meaning a reasoned account. Physiology thus is the science of how living things function. Plant physiology is an exciting discipline that has advanced spectacularly since World War II. The war effort accelerated the development of instrumentation, produced radioactive compounds that helped trace physiological processes, and fostered an intellectual and financial environment conducive to scientific inquiry.

Figure 1.1 Collecting sap, which can be consumed directly or fermented to produce pulque, from the maguey *Agave salmiana*, during the Aztec period. Detail is from a mural in a series by Diego Rivera at the National Palace in Mexico City.

Such technical advances and the knowledge they generated have set the stage for understanding the promising future for agaves and cacti. The historical success of these plants is well documented. For instance, the most successful crop in eastern Africa—now Kenya and Tanzania—in the early part of the twentieth century was sisal (*Agave sisalana*). However, some species have been too successful; imported prickly pear cacti spread along the eastern coast of Australia with devastating ecological and agronomic effects. This unfortunate Australian experience underscores the competitiveness and productivity of these plants. In particular, under the right conditions, certain agaves and cacti can produce more biomass than can essentially all other cultivated plants (plants that have been carefully bred for productivity over millennia). To appreciate the high productivity of certain agaves and cacti is relatively easy, but to understand the physiological reasons for this productivity is more difficult but well worth the effort.

Classification and Evolution

As we begin our examination of agaves and cacti, we must reckon with the great differences between the two groups. Biologists tend to divide things into mutually exclusive categories, a process that helps classification but requires a huge scientific vocabulary. For instance, is something animate or inanimate, plant or animal, monocotyledonous or dicotyledonous? Although the first two pairings are relatively clear, the third is more technical and related to the two categories of the flowering plants (Angiospermae).

Grasses, lilies, and palms are *monocotyledons*, and so they initially produce a single leaf axis and have leaves with parallel veins, as do agaves. Most flowering plants, such as cacti, ivy, lettuce, and oaks, are *dicotyledons*. Dicotyledons initially produce two leaf axes and have leaf veins with a netlike pattern, stems that grow progressively thicker with age, and roots that tend to branch. Taxonomically speaking, we could hardly pick two groups of flowering plants that differ more from each other than do agaves and cacti. Yet physiologically speaking, these two *taxa* (groups) have converged on many of the same solutions to the problems of growing in regions of low rainfall. These regions are often referred to as *arid* (corresponding approximately to regions with less than 250 millimeters [10 inches] of annual rainfall) and *semiarid* (250 to 450 millimeters [10 to 18 inches] of annual rainfall). Arid and semiarid regions comprise about one-quarter of the land area of North America plus South America, where agaves and cacti are native, and about one-third of the land area of the whole earth.

The plant kingdom contains about 300,000 species, ranging from mosses and ferns, through coniferous plants such as pines, to the flowering plants. Flowering plants represent the summit of plant evolution, and angiosperms make up the dominant vegetation on the earth. About 420 families of flowering plants are recognized, of which 70 are monocotyledons and 350 are dicotyledons. Each plant family is divided into genera (singular, genus), which are further divided into species. The genus Agave, which means noble (from the Greek *agauē*), was first described by the Swedish botanist and father of modern taxonomy, Carolus Linnaeus. In 1753, he named its first species, *Agave americana* (Figure 1.2). Agave is the main genus in the monocotyledonous family Agavaceae and has 136 species. Cactus, a genus also described by Linnaeus, is the common name for the dicotyledonous family Cactaceae, which has about 1600 species in 122 genera. The word *cactus* derives from the Greek *kaktos*, referring to an unrelated prickly plant found in southern Europe and northern Africa.

Beginning with a hypothetical ancestor of the angiosperms about 200 million years ago, plants have undergone genetic changes leading to the divergence of various lines into recognizable groups. The events separating the monocotyledons from the dicotyledons probably occurred about 140 million years ago. The progenitors of modern plants kept diverging over millions of years, leading to an ever-accumulating set of differences. Grass, a monocot, now has a very different morphology, a different pattern of growth, and a different method of reproduction than does ivy, a dicot.

The line that led to agaves apparently originated about 60 million

Figure 1.2   Agave americana, which is planted around the world as an ornamental, was the first species placed in the genus Agave.

years ago. Climatic changes leading to the spreading of this taxon into new regions and to its *radiation* (the evolution of new species) took place more recently, probably 30 to 35 million years ago. Judging from the fossil record as well as the distribution in nature of present-day agaves, we can pinpoint the initial site of such radiation in what is now southern Mexico or northern South America. One might be tempted to say that the origin was in Central America, but the shift in tectonic plates near the earth's surface had not yet created Central America at that time! Based on flower structure, tissue properties, and specific chemical compounds, close taxonomic relatives of the agaves include amaryllis and onion (family Amaryllidaceae; some taxonomists place onions in their own family, Alliaceae) and asparagus (family Asparagaceae).

The evolutionary divergence of cacti from other flowering plants probably occurred 70 to 90 million years ago, again in the New World, and again the major radiation of new species probably occurred 30 to 35 million years ago. The center of origin of this family is similar to that of the agaves. The closest relatives of cacti are most likely in the families Portulacaceae, which contains the portulacas, and Didiereaceae, which contains the didiereas and is confined to Madagascar.

Although present-day agaves are apparently rather homogeneous genetically, cacti show great diversity. For instance, all agaves have prominent leaves and single vertical stems that are usually hidden by the leaves. On the other hand, cacti can range from primitive species with large leaves and thin stems (which can almost be mistaken for ivy, such as many species of *Pereskia*) to rather short shrubs with readily detachable stem segments (examples include the *cylindropuntia* with cylindrical stems [*Opuntia bigelovii*] and the *platyopuntia* with flattened stem segments [*Opuntia ficus-indica*, a prickly pear cactus]) to tall treelike plants with prominent columnar trunks (such as the saguaro [*Carnegiea gigantea*], which can be found in the Sonoran Desert of United States and Mexico, and *Trichocereus chilensis*, which is widely distributed in Chile).

These three sets of examples are representative of the three subfamilies of the Cactaceae: subfamily Pereskioideae (about 20 species), all of which have prominent leaves; subfamily Opuntioideae (200 to 250 species), about 70 percent of which are in the large and diverse genus *Opuntia* and have small deciduous leaves; and subfamily Cactoideae, the largest and essentially leafless subfamily containing 1300 species. The Cactoideae are divided into nine tribes, with *tribe* being the taxonomic ranking just below a subfamily. The main genera of subfamily Cactoideae in terms of numbers of species are *Mammillaria* (about 200 species in tribe Cacteae); *Echinocereus* (about 50 species in tribe Echinocereeae); *Gymnocalycium, Neoporteria, Parodia,* and *Rhipsalis* (about 40 to 60 species each in tribe Notocacteae); and *Melocactus* (about 35 species in tribe Cereeae).

Distribution: Threatened Species

Based on the time and the site of their evolutionary origin, agaves and cacti might be expected to occur naturally only in the New World, which is essentially the case. The greatest concentration of native species of agaves and cacti occurs in the southern third of North America; liberally throughout Central America, including the Caribbean islands; and in the northern half of South America. Species abundance drops off northward and southward, but the distributions of some cacti extend into Canada to the north and into southern Argentina and southern Chile to the south.

Those who have seen the tall leafless succulents of southern Africa may have sworn that they were seeing large columnar cacti. However, these leafless trees, which can dominate the landscape in Zimbabwe, Madagascar, and South Africa, are in an entirely different family, Euphorbiaceae. Unrelated plants cope with similar environments in geographically different habitats by adopting similar growth forms, a process known as *convergent evolution*. The columnar shape and the leaflessness of members of both the Euphorbiaceae and the Cactaceae are examples of convergent evolution that has taken place over long periods of time between unrelated plants. Although the cylindrical succulents of Africa are not cacti, some species of the primitive cactus genus

Rhipsalis are apparently native to Madagascar and to a few other parts of the Old World.

The present-day continents of Africa and South America began drifting apart about 125 million years ago, long before the Cactaceae had originated and radiated. Some unknowing bird or perhaps a fortuitous piece of driftwood may have been involved in spreading from the New World the only genus of agaves or cacti that is native to the Old World, *Rhipsalis. Rhipsalis baccifera* (Figure 1.3), which is native to North and South America, has fruits about 6 millimeters (0.25 inch) in diameter that are irresistible to certain birds. Birds can spread the seeds via their feces and hence could have expanded the distribution of this species over great distances, perhaps eventually to Africa. Indeed, *R. baccifera* can be found in Kenya and Sri Lanka, and two other species of *Rhipsalis* occur in Madagascar.

The beauty of many species of agaves and especially cacti threatens their existence in the wild. When the commercial market craves a diminutive cactus with a beautiful flower that happens to grow only on some mountainside, how do we balance the economic opportunity against a permanent ecological loss? Although the voices of conservationists have grown louder over the past decade, financial incentives have unfortunately caused collectable species to become locally extinct and have even threatened the loss of entire species. The long-range consequences of such short-sighted behavior can be bemoaned but are nonetheless rather common. For instance, the picturesque and characteristic barrel cacti of Arizona, *Ferocactus covillei* and *F. wislizenii*, are now rarely found within 50 meters (160 feet) of roads, in contrast with their distribution pattern in the early twentieth century. Old distribution maps are a ghostly indication that our collection ability exceeds the rate of natural reproduction. Moreover, habitat destruction accompanies the ever-growing human population, further reducing the region where such cacti can survive.

To help eliminate the ecologically devastating trade in certain native plants, and hence to prevent or at least to postpone extinction, restrictions have been placed on the removal of certain plants from their natural habitat and especially on their movement between countries. These restrictions are enforced by government officials at international borders. The United States Endangered Species Act of 1973 defines *endangered* as "any species which is in danger of extinction throughout all or a significant portion of its range" and the lesser category of *threatened* as "any species which is likely to become an endangered species within the foreseeable future throughout all or a significant portion of its range." The passage of such federal legislation has encouraged states to enact similar legislation protecting species within their own boundaries. About 20 species of cacti—more species than for any other plant family—have

Figure 1.3   Rhipsalis baccifera, an example of the only genus of agaves or cacti that is native to the Old World. Note the many fruits, whose seeds can be readily dispersed by birds.

been placed on the threatened and endangered species list for the United States. Certain varieties of the cacti *Coryphantha vivipara, Echinocereus engelmannii, Opuntia basilaris, O. munzii, O. parryi,* and *O. wigginsii* were or are now on the threatened and endangered species lists for California. Among agaves, *Agave utahensis* is under federal surveillance for possible future inclusion on such lists; *Agave shawii* is being so considered in California; and *Agave arizonica* is already on the U.S. endangered species list.

As an outcome of the increasing public awareness of the long-term value of native plants, Arizona law now stipulates that native cacti must be tagged and a fee paid before they can be moved. California has set up strict requirements for the movement of native succulents, including essentially all agaves and cacti. The affirmation on a U.S. Customs Declaration form that one is transporting an agave or a cactus into the United States usually guarantees a conversation with a representative of the U.S. Department of Agriculture concerning CITES (Convention on International Trade in Endangered Species of Wild Fauna and Flora). Under CITES regulations, the movement of plants across international boundaries requires appropriate permits and documentation, which must be provided for *Agave arizonica, A. parviflora, A. victoriae-reginae*, and all cacti except some used agronomically. To enforce the CITES convention, the U.S. Fish and Wildlife Service of the Department of the Interior uses Latin binomials, not common names, for plant identification. The reason is that common names are imprecise and vary from location to location. At least 10 species of agave are referred to as "century plant," and over 20 species of cacti are called "prickly pear." We will use both scientific and common names, favoring common names when they are widely used and not ambiguous.

Features in Native Environments

Agaves and cacti have many unique adaptations to their native environments, including several strategies for reproduction. Most species reproduce by means of flowers, fruit, and seeds—the conventional way of leading to future generations. For some species, propagation is vegetative, by a part of the plant that is genetically identical to the parent plant. We will also consider two other interesting aspects, sexual versus vegetative reproduction as it relates to the average age of a common agave, and the tilting of the stems of certain barrel cacti toward the equator.

Pollination

For flowering plants to reproduce, pollen grains must be moved from the stamens of a flower to the stigma of the same or a different flower (Figure 1.4A). Pollen grains (Figure 1.4B and C) are about 0.04 millimeter (0.002 inch) across and so are just barely discernible by the human eye. Their small size permits them to be transported by the wind, but wind pollination is rather inefficient and probably not important for agaves or cacti. Because one pollen grain is needed for each fertile seed, and the fruits of agaves and cacti generally contain many seeds, efficient

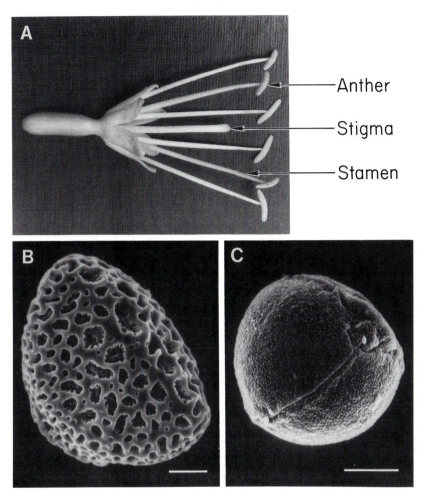

Figure 1.4 Reproductive structures: (A) flower of *Agave vilmoriniana*, showing the pollen-containing anthers on the stamens and the pollen-receiving stigma. Scanning electron micrographs of pollen from (B) the agave *Agave sobria* variety *roseana* and (C) the cactus *Parodia nivosa*. The scale bars indicate 10 micrometers (0.0004 inch). (Photographs were prepared by Gretchen B. North and Bingru Huang.)

and specific pollen transfer is necessary for the sexual reproduction of these succulents. Pollen is transported by animals ranging from tiny insects to domestic honeybees to bats. Agaves and cacti tend to require *cross-pollination* from one plant to another by such animals for successful seed production. Moreover, cross-pollination helps maintain the genetic diversity that is crucial to evolutionary change.

Bat pollination is common for many species of agaves and some species of cacti that have relatively large flowers producing nectar with a pungent, rather unpleasant odor, such as *Agave palmeri* and the cactus *Stenocereus thurberi* growing in Arizona. The flowers of many agaves and cacti are long and tubular, often an indication that hummingbirds are the pollinators. Because humans influence the range of hummingbirds by providing feeders containing sugar solutions, humans can change the pollinators for native species. An even greater impact on the pollination of native species is caused by domestic bees raised for honey production. The sheer number of such honeybees in certain regions ensures that they will play a major role in pollination, thereby changing the importance of native pollinators. The flowers of a cactus such as senita (*Lophocereus schottii*) open at night and are pollinated by nocturnal moths. Many species of native bees and wasps can be seen visiting the flowers of other cacti, including the tallest cactus in the United States, saguaro (*Carnegiea gigantea*). In fact, bees pollinate 80 to 90 percent of agaves and cacti. Hawkmoths are also important pollinators, but the importance of ants and beetles for the pollination of agaves and cacti is subject to debate.

One of the keys to the successful development of new cultivars of cacti has been pollination by hand. This is exceedingly simple. The pollen may be teased out of the anthers with a small paintbrush, a cotton swab, or even the body of a dead honeybee. The pollen is then applied to the stigma (Figure 1.4A) of the flower to be fertilized. Also, stamens (Figure 1.4A) may be plucked out with tweezers and the pollen merely knocked out of the anthers onto a stigma. What happens next depends on providence and the talent of the breeder. Successful hand pollination is required to satisfy the huge demand for ornamental cacti with interesting growth forms and beautiful flowers. These plants are increasingly being raised from seed in large commercial nurseries worldwide.

Vegetative Propagation

Lest one feel that sex is always crucial for agaves and cacti, we note that most of their agronomically important species are not propagated by seed. The sisal agave (*Agave sisalana*), cultivated throughout the twentieth century in plantations in eastern Africa, has been propagated by bulbils. *Bulbils* are young plantlets that occur on the *inflorescence* (flowering stalk). The bulbils of *A. sisalana* are generally at least 10 centimeters (4 inches) in length when harvested and can be placed directly in the soil. This curious means of reproduction is also used by various other species of agave, such as *Agave angustifolia* from central Mexico, which is cultivated for beverage production.

Large prickly pear cacti, such as *Opuntia ficus-indica*, are propagated by removing individual flattened stem segments, termed *cladodes*, when they are about 20 centimeters (8 inches) or more in length. The cladodes are deliberately allowed to dry for a few days, often in full sunlight, to discourage bacterial and fungal infections. Although this procedure would kill the stems of nearly any other plant, for these cacti it merely allows calluses to form on the cut surfaces. The cladodes are then placed vertically in the ground with about one-quarter of their surface area in the soil. This easy and basically foolproof technique of propagation offers a significant head start on growth compared with planting a tiny seed. Many cacti are propagated by grafting a portion of the stem of one species onto a stem of the same species or even that of a distant cousin. Grafting often leads to attractive—if bizarre—ornamental plants.

Most older agaves produce multiple underground stems, termed *rhizomes*, from the base of their stems. A new plant, or *ramet*, emerges from the end of the rhizome. Such ramets are initially dependent on the mother plant. Nearly all of the reproduction of the relatively slow-growing common desert agave (*Agave deserti*; Figure 1.5), which is native to the Sonoran Desert of the southwestern United States and northwestern Mexico, occurs by such vegetative means. When a mature agave established by seed eventually flowers and dies, its descendants produced on rhizomes will have radiated out from its base. Subsequent

Figure 1.5 Large oval ring of *Agave deserti* that propagated vegetatively after the initial establishment of a seedling hundreds of years ago. The living plants are on the periphery of the oval, with the original plant and its ramets forming a mesh of dead material at the center of the oval. The site is Agave Hill, at an elevation of 850 meters (2800 feet) above sea level in the University of California Philip L. Boyd Deep Canyon Desert Research Center just south of Palm Desert, California.

rhizomes on the descendant plants cause a ring of vegetatively produced plants to expand with time (Figure 1.5). This creates a growth pattern similar to that of the famous "fairy rings" of various mushrooms, which can be thousands of years old. The large ringlike groupings observed for *A. deserti* and other agaves in the field may also have long lifetimes.

Age and Propagation of Agave deserti

Individuals of *Agave deserti* flower only once and then die, a life history generally referred to as *monocarpy*, which characterizes three-quarters of the species in its genus. The percentage of the plants of *A. deserti* that flower in a particular year in the southwestern United States varies from a low of about 0.1 percent to a high of nearly 4 percent. Years with a low flowering percentage alternate with years with a high flowering percentage, just as for various fruit trees and many other perennials. An average of 1.9 percent of *A. deserti* flower in a particular year, so the fraction of the plants that flower averages 0.019 a year. The average number of years to reach flowering age, which is the *mean lifetime* for monocarpic plants, equals the reciprocal of the average number of plants flowering in a year. For example, if an average of one-fifth of the plants flower each year, then the time for an individual plant to flower averages 5 years. Thus the mean lifetime for *A. deserti* is $1/(0.019 \text{ year}^{-1})$, or 53 years. Fifty-three years, however, is much shorter than the lifetime expected based on the name "century plant," which is often applied to this species (as well as to other species of agaves).

Seedling establishment is extremely rare for *A. deserti*, even though each inflorescence can produce about 65,000 seeds. Over a 29-year period at one field locale, only 2 years (1967 and 1982) led not only to seeds that germinated, which occurred basically every year, but also to seedlings that survived the droughts of their first year. Droughts exact an extremely high toll on young seedlings. Their small size causes them to shrivel and die before the next rainfall provides the moisture needed for rejuvenation and survival, a problem not encountered by older larger plants with more stored water. Thus only about 1 percent of the propagation of *A. deserti* is by seedlings and 99 percent is by ramets. The ramets, which may depend on the original rosette for water and sugars for up to 15 years, are produced every year, with the number varying with the favorability of a particular year for growth. Such favorability can be expressed based on several environmental factors, especially water availability (the main factor limiting the growth of desert succulents), air temperature, and light.

Tilting of Barrel Cacti

The stems of certain barrel-shaped cacti of the southwestern United States tend to point southward toward the equator. *Ferocactus covillei* and

F. wislizenii in Arizona hence were referred to as "compass plants" by the miners and others traveling through such regions in the nineteenth century. Barrel cacti in the Southern Hemisphere, including various species of *Copiapoa* in northern Chile (Figure 1.6), also tilt toward the equator. Although the importance of distinguishing north from south for humans is obvious, the ecological consequences of such stem tilting is less so. But plants that orient toward available light tend to grow

Figure 1.6 Northward equatorial tilt of *Copiapoa cinerea* near Pan de Azucar, Chile.

better, so pointing away from competing vegetation and toward the equator may be advantageous for light interception. In addition, such an orientation raises the temperature at the top of the stem, which can hasten and increase flowering.

Human Uses

Nine-thousand-year-old mummified human feces found in caves in Mexico contain fragments of agaves and cacti! The collection of fruit from many species of cacti, formerly by Indians over wide areas in the present-day southwestern United States and northwestern Mexico but now in much more restricted areas, can alter the abundance of such plants. Similarly, the age-old practice of harvesting the inflorescences of various agaves, whose native habitats are being reduced in size by conversion into cultivated land, can also affect the diversity of ecosystems. How do we balance the traditional uses by indigenous cultures with the increasing interest in preserving genetic diversity?

We might next ask why agaves and cacti have been so important to indigenous cultures. Let us start with the reproductive organs, which include conspicuous flowers and tantalizingly sweet fruit. Why not plant such species near homes, or at least develop pathways to their natural habitats? Both appear to have been done for many species of cacti. The leaves of agaves last for many years, and so a logical conclusion is that they would make an ideal thatch for a roof. People also discovered that the fibers of certain agaves are strong, suggesting that they could be woven into ropes and textiles. Sugar-containing liquids from agaves may have first been harvested accidentally. A wound near the base of an agave stem may have filled up overnight with a sweet elixir, a potion that changed naturally within a few days to an alcohol-containing beverage for a different clientele. Humans also probably noticed that platyopuntias were eaten by wild animals, ranging from rodents and rabbits to javelinas (peccaries) and deer, and the aridity of the land during a drought may have driven both humans and their livestock to do likewise. This helped lead to the many ways that young cladodes are currently eaten—blanched, marinated, or cooked—as well as to over 600,000 hectares of platyopuntias that are now planted worldwide for cattle fodder (1 hectare = 10,000 meter^2 = 2.47 acres).

We can supplement our speculations on the prehistoric uses of agaves and cacti with ethnobotanical records of the Seri Indians of northwestern Mexico (Table 1.1). These highly independent native Americans, who numbered just under 200 in the 1930s, have their own language and a culture in tune with their desert environment. The richness of their knowledge of agaves and cacti is unsurpassed. They harvest the fruits of about 20 species of cacti, mostly to be eaten raw, and they also dry and store fruits and seeds for future use. They use fluids from cacti as an emergency water ration, as a sweet drink, and, after

Table 1.1 Uses of Cacti by Seri Indians in Northwestern Mexico

Use	Species	Comment
Fruit	Barrel cacti (*Ferocactus acanthodes, F. covillei,* and *F. wislizenii*), cardón (*Pachycereus pringlei*), *Cereus striata,* chain-fruit cholla (*Opuntia fulgida*), desert Christmas cactus (*Opuntia leptocaulis*), Engelmann prickly pear (*Opuntia phaeacantha*), fishhook cactus (*Mammillaria estebanensis* and *M. microcarpa*), golden hedgehog cactus (*Echinocereus engelmannii*), organ pipe (*Stenocereus thurberi*), pencil cholla (*Opuntia arbuscula* and *O. versicolor*), pitaya (or pitahaya) agria (*Stenocereus gummosus*), purple prickly pear (*Opuntia violacea*), saguaro (*Carnegiea gigantea*), senita (*Lophocereus schottii*), sina (*Stenocereus alamosensis*)	Cactus fruit was probably the most important single food source for the Seri Indians as well as their favorite plant food. It could be stored after drying; the seeds were often removed and eaten separately, sometimes after roasting (barrel cacti, cardón, saguaro); and the flower petals were also occasionally eaten (especially those of barrel cacti).
Food (other than fruit)	Many, including teddy bear cholla (*Opuntia bigelovii*) and chain-fruit cholla	Stems were eaten, often after roasting, and a gum exuded from the stem was eaten raw or after cooking.
Wine	Cardón, organ pipe, pitaya agria, saguaro	Mashed fruit was fermented for a few days. When water was added, the resulting wine had a lower alcohol content.
Water	Barrel cacti (especially *F. wislizenii*)	Liquid was squeezed from the pulp (though its consumption often caused muscle pain and diarrhea).
Containers	Barrel cacti, cardón, saguaro	Spines were burned off hollowed stems (barrel cacti, especially *F. wislizenii*). The callus tissue lining a woodpecker nest remained after the stems decayed (cardón, saguaro).

(*continued*)

Table 1.1   (*Continued*)

Use	Species	Comment
Poles	Cardón, organ pipe, saguaro	"Ribs" from wood were used for prying off fruit, as handles of drills and other tools, for constructing shelters, as torches, and for smoke signaling (organ pipe).
Games	Organ pipe, saguaro, senita	Fruit skins were used in a game like darts (saguaro), and stem segments with spines removed were part of a game like dodgeball.
Toys	Organ pipe, senita	Cylindrical stem segments with their central fleshy parts removed left wheellike disks at either end.
Medicine	Barrel cacti, cardón, *Cereus striata*, chain-fruit cholla, fishhook cactus, *Opuntia marenae*, organ pipe, saguaro, teddy bear cholla	Besides being used to relieve general aches and pains and even rheumatism (saguaro), tuberlike roots were prepared as a remedy for swelling (*Cereus striata*); juice from cooked stems was used as a remedy for earache (fishhook cactus); and the root was used as diuretic (teddy bear cholla).
Sealant	Saguaro	Juices were obtained from flowers.
Caulking	Organ pipe, pitaya agria	Pitch made from stems was used on boat seams and on pottery.
Ceremony	Organ pipe	Stem sections were used in girls' puberty ceremony.
Religion, ritual	Barrel cacti, cardón, organ pipe, saguaro, senita, teddy bear cholla	Stems were used for burial, creating or stopping rainfall, stopping the wind (organ pipe), long life, placing curses (senita), and burning a dead girl's property (teddy bear cholla).

Table 1.1 (*Continued*)

Use	Species	Comment
Tanning	Cardón, saguaro	Extracts were obtained from seeds.
Tattoo	Cardón	Inks were obtained from fruit.
Face paint	Barrel cacti, Englemann prickly pear, purple prickly pear	Extracts were used from young spines, flowers (barrel cacti), and fruit (prickly pear cacti).
Fishing, hunting	Cardón, golden hedgehog, organ pipe	Blinds were fashioned for hunting, and stems were also used for securing carcasses above the ground (cardón, organ pipe). Inner tissues were used as fish lures (golden hedgehog).

Note: For more details, see Felger and Moser (1985).

fermentation, as an alcoholic beverage. They make myriad products from cacti, including containers, games and toys, sealants for clay vessels and boats, face paint, and tatoo ink. The Seri Indians even use products from cacti as medicine and in religious observances (Table 1.1).

In addition to cacti, nine species of agaves have at one time been part of the day-to-day lives of Seri Indians. Of special importance are the stems, the bases of leaves clustered around them, and particularly the inflorescences, all of which are generally roasted before being eaten. The stems and other plant parts are usually roasted in large earthen pits on glowing embers and then covered with flat rocks on which another fire is made. The entire assembly is covered with soil to permit overnight baking. The roasted tissue is prepared in various ways or is preserved by drying. *Agave cerulata* is used as an emergency water source or to make wine, and bundles of its leaf fibers are used as brushes. Wine is also made from crushed stems of *Agave colorata*. *Agave schottii* yields a foamy shampoo from its crushed leaves and arrow shafts from its inflorescences. Necklaces are created from the flat black seeds of *Agave subsimplex*, face paint from its cooked stems, and poles for picking cactus fruit from its inflorescence.

Human interest in agaves and cacti continues unabated, with more than 10 periodicals devoted to such plants. Amateur societies have sprung up around the world to share in the pleasures and mysteries of cacti and other succulent plants. Membership in these societies or subscribers to such periodicals currently number about 12,000 in the two republics that were Czechoslovakia, 10,000 in Germany, 7000 in Great Britain, 10,000 in Japan, and 7000 in the United States. We will examine the many uses of agaves and cacti, most of which are unappreciated outside Latin America, in Chapters 2 and 3. The productivity of agaves and cacti (Chapter 7), including the possibilities of increased utilization with intelligent management and foresight (Chapter 8), suggests an ever-increasing role for these plants in the future.

Morphology and Anatomy

Most people can identify a cactus based on its morphology, but a surprising number call an agave a cactus. This is forgivable physiologically, as we shall see, but hard to understand morphologically. Agaves have thick, pointed leaves that unfold from a central spike of folded leaves, whereas the preeminent feature of the shoots of most cacti is their lack of leaves. The stems of cacti can be rather cylindrical, as for the barrel cactus *Ferocactus acanthodes* (Figure 1.7A) and the saguaro (*Carnegiea gigantea*; Figure 1.7B), or they may consist of a series of stem segments, as for the cladodes of the prickly pear cactus *Opuntia ficus-indica* (Figure 1.7C). Another prominent and unique feature of cacti is the *areole*, a regularly occurring stem region where cells can divide to produce spines and other structures. For instance, the nasty deciduous *glochids* are produced by the areoles of opuntias—nasty because these tiny spines contain barbs that can progressively work their way into your skin. Even if you only gently brush the areoles, the glochids become detached and lodge in your skin, a painful experience familiar to anyone dealing with platyopuntias.

Viewed from the top, the angle of a newly unfolding leaf of an agave is almost exactly 137° from the previously unfolded leaf, either clockwise or counterclockwise around the spiral of leaves on the stem. Such a pattern can be "interpreted" using the famous numerical sequence identified by the Italian mathematician Leonardo Fibonacci in about 1200. Any number in the Fibonacci sequence is the sum of the two numbers before it: 1, 1, 2, 3, 5, 8, 13, 21, 34, and so forth. How this sequence of numbers relates to plants is almost magical. The angle between two successively unfolding leaves can be thought of as a fraction of a circle, a

---

→

Figure 1.7 Stem morphology of various cacti: (A) top view of *Ferocactus acanthodes* having 21 ribs, a Fibonacci number, at the field site in Figure 1.5; its spines have been cut away from the apex, revealing a layer of hair, or pubescence, that covers the growing region known as the apical meristem; (B) a forest of *Carnegiea gigantea* in Saguaro National Monument, Arizona; and (C) *Opuntia ficus-indica* used as a hedge in Israel; its fruits are harvested and sold in informal markets along the roadsides.

fraction that is the ratio of two Fibonacci numbers. These two numbers can be determined by counting leaves around the stem of an agave. The first number is the number of spirals or turns around the stem that must be traversed before a leaf occurs in exactly the same orientation as the starting leaf, and the second number is the number of leaves that were passed while counting the number of spirals. Starting with any leaf of an agave, 21 leaves must unfold in 8 spirals around the shoot axis before one points in exactly the same direction as the original leaf. The ratio of these Fibonacci numbers, 8 divided by 21, times 360° (the number of degrees in a circle) is 137°, the observed angle between successively unfolding leaves of agaves.

The intriguing Fibonacci sequence also extends to the number of ribs of certain barrel-shaped cacti. For example, a common barrel cactus of the southwest United States, *Ferocactus acanthodes*, generally has 5 ribs when it is 5 centimeters (2 inches) tall, 8 ribs at 8 centimeters in height, 13 ribs at 15 centimeters, and 21 ribs at about 25 centimeters in height (Figure 1.7A). The ribs are produced in pulses of Fibonacci numbers in the region of new growth at the top of the stem, the *apical meristem* (a region involved with the formation of new cells).

The sizes of agaves and cacti have important physiological conse-quences that can be expressed in terms of the ratio of the shoot volume to the shoot surface area. As an example of a volume:area ratio, let us consider an approximately spherical shoot, such as that of a young barrel cactus. The volume of a sphere equals $^4/_3\ \pi r^3$ and its surface area equals $4\pi r^2$, where r is the radius. Hence the volume in which water can be stored per unit surface area across which water can be lost is $(^4/_3\ \pi r^3)\ /\ (4\pi r^2)$, or $r/3$. As a barrel cactus increases in size, its volume thus increases more rapidly than does its surface area. A larger spherical cactus can therefore better withstand drought than a smaller one can, because the water-storage volume per water-loss area is greater for the larger cactus. This concept of volume:area ratio has important conse-quences for small seedlings, which have low values of $r/3$ and hence a very limited ability to tolerate drought. The most vulnerable time for seedling survival in the field is therefore their first drought, as already mentioned for *Agave deserti*.

Roots have received much less research attention than have shoots, primarily because direct observation of roots is generally not possible. Roots are required for water and nutrient uptake and are relatively shallow for agaves and cacti. These plants usually are found in sandy soils, which facilitates root excavations. Root distributions are conse-quently better known for agaves and cacti than for most other peren-nials. The main roots of agaves tend to be thin and to radiate out from the base of the stem without branching, as is typical of most mono-cotyledons; fine lateral branches grow from the main roots and are usually shed during drought. The roots of cacti, on the other hand, tend

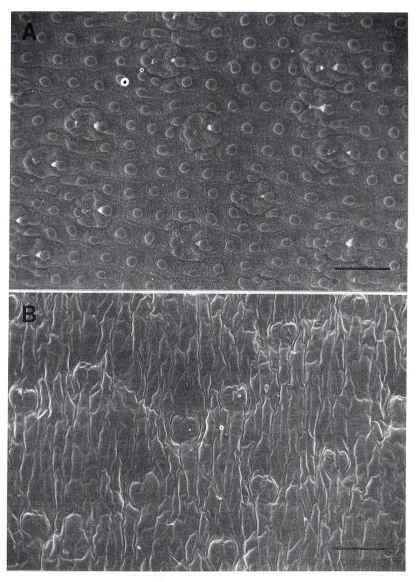

Figure 1.8 Surfaces of the shoot, illustrating stomatal pores: (A) leaf of *Agave attenuata*, whose stomates tend to occur in rows, as for other monocotyledons; and (B) stem of *Opuntia ficus-indica*, whose stomates occur in the random pattern typical of dicotyledons. Scale bars indicate 0.1 millimeter (.004 inch). (Photographs were prepared by Bingru Huang.)

to branch and rebranch as well as to grow thicker with age, both characteristics of dicotyledons.

As do other terrestrial plants, agaves and cacti lose water primarily through pores between special cells in their leaf and stem surfaces (Figure 1.8). These pores are called *stomates* (or *stomata*) from the Greek *stoma*, meaning mouth. Such "mouths" regulate the entry and exit of gases for the shoot and are enclosed by a pair of cells referred to as *guard cells*. When the guard cells swell by taking up water, the pore is opened. Stomatal opening greatly enhances the movement of gases between the plant shoot and its environment. Conversely, stomatal closure restricts the exchange of gases with the environment and generally occurs during the night for most plants, but not for agaves and cacti. The outermost layer of cells of the shoot, the *epidermis*, is covered by a waxy waterproofing layer, the *cuticle*, which helps prevent water loss from the plants to the surrounding air.

Growing plants take up carbon dioxide (CO_2) from the atmosphere. The CO_2 enters a plant when its stomates are open. The process of photosynthesis—during which CO_2 is incorporated into usable products, such as the sugar glucose—takes place in the *chloroplasts*. Chloroplasts, which individually appear pale green when examined with a light microscope because they contain the green pigment chlorophyll, occur in cells that are just under the shoot epidermis. When the shoots are cut open with a knife or a razor blade (Plate B), the outermost 1 to 5 millimeters (0.04 to 0.2 inch) appear green because of chlorophyll. Beneath this *chlorenchyma* is a tissue with whitish cells that generally lack chloroplasts, the *water-storage parenchyma*, which is involved in water storage (Plate B). The chlorenchyma is involved in CO_2 uptake, and the water-storage parenchyma is crucial for the drought tolerance of agaves and cacti.

Physiology: Water Conservation

The ecological success of agaves and cacti relates to their unique way of breathing—how they exchange oxygen, carbon dioxide, and water vapor with the environment. When the sun comes up, most plants open their stomates and take up CO_2 from the atmosphere. But an inevitable consequence of the opening of stomates is the loss of water from the inside of the leaves and stems. What if the overriding consideration were to save water? Is stomatal opening during the daytime then the best strategy?

To facilitate the exchange of gases between a shoot and the surrounding air, the stomates must be open. Gas molecules enter or leave the shoots of agaves and cacti by moving toward regions of lower concentration, a process called *diffusion*. The rate of diffusion increases as the concentration difference between the two regions increases (Figure

A   Daytime

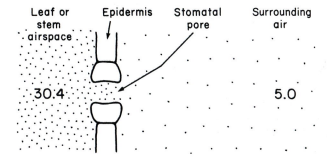

B   Night

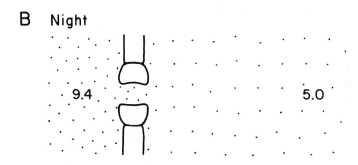

Figure 1.9   Schematic illustration of the water vapor concentrations in the air spaces of the chlorenchyma of a leaf or a stem and in the surrounding air: (A) during the daytime at 30°C; and (B) during the night at 10°C. The much higher water vapor concentration difference from the shoot to the surrounding air during the daytime leads, for the same stomatal opening, to a higher rate of diffusion of water vapor out of the plant than at night. The numbers indicated are the water vapor concentrations in grams meter^{-3}, as discussed in the text.

1.9). Water vapor inside the leaves and stems of plants is essentially at the saturation concentration, which increases with temperature. For instance, the mass of water vapor per volume of saturated air at atmospheric pressure is 9.4 grams meter^{-3} at 10°C (50°F), 17.3 grams meter^{-3} at 20°C (68°F), and 30.4 grams meter^{-3} at 30°C (86°F). Thus the concentration of water vapor inside agaves and cacti nearly doubles for each 10°C increase in air temperature (gram meter^{-3} is a typical unit for water vapor concentration, where meter^{-3} means "per meter^3" or "per cubic meter"; at atmospheric pressure and 20°C, dry air has a mass of 1205 grams meter^{-3}).

In a Mediterranean climate, such as that in southern Italy and Greece, coastal southern California, and west-central Chile, the air might contain a water vapor concentration of 5.0 grams meter^{-3} (Figure

1.9). On a warm summer day with a temperature of 30°C, a water vapor concentration of 5.0 grams meter^{-3} in the surrounding air corresponds to a relative humidity of 17 percent. If the stomates were to open during the daytime when the shoot temperature is also 30°C, the difference in water vapor concentration from the shoot to the surrounding air would be 30.4 minus 5.0, or 25.4 grams meter^{-3} (Figure 1.9A). If the stomates were to open at night when the shoot and air temperatures are 10°C, the concentration difference leading to water vapor diffusion out of the shoots would be 9.4 minus 5.0, or only 4.4 grams meter^{-3} (Figure 1.9B; in air at 10°C, a water vapor concentration of 5.0 grams meter^{-3} corresponds to a relative humidity of 53 percent). Thus the water vapor concentration differences leading to water diffusion out of a shoot, a process called *transpiration*, is six times (25.4 divided by 4.4) higher during the daytime at 30°C than during the nighttime at 10°C. A plant would consequently lose six times more water for stomatal opening during the daytime, compared with the same stomatal opening at night.

Let us consider this conclusion qualitatively because it is crucial to understanding the advantages that agaves and cacti have over most other plants for water conservation. We will use a simple analogy to help explain the water-conserving advantage of opening the stomates at night. Imagine trying to dry clothes or a wet towel outdoors during the heat of the day versus the cool of the night. The process of losing water from the wet clothes requires the diffusion of water vapor from the clothes to the surrounding air. The water vapor content in the pores between the fibers is basically the saturated value at that temperature, as is also the case inside plant shoots. A wet towel will dry in perhaps an hour outside at noon on a sunny day but will remain wet if the drying were foolishly attempted at night—foolish for drying clothes, that is, but extremely advantageous for conserving water in plants.

The analogy of drying clothes at night versus during the daytime can help illustrate the advantage of opening stomates at night for limiting water loss. But a legitimate and difficult question is how to store carbon dioxide taken up at night. Without a net CO_2 uptake, a plant gradually loses its sugar reserves and eventually dies. Nearly all plants take up CO_2 during the daytime, when the energy of sunlight is directly available. But agaves and cacti take up CO_2 at night, for which the biochemistry will be explained in Chapter 6 and applied to biomass productivity in Chapter 7. The clear advantage—conservation of water—involves some special biochemical quirks for taking up and storing CO_2 at night.

The strange practice of nighttime CO_2 uptake for certain plants was stumbled on in the early part of the nineteenth century. While Benjamin Heyne, a Moravian missionary, physician, and accomplished amateur botanist, was in India in 1813, he took bites out of *Kalanchoe pinnata* at various times of the day. *Kalanchoe pinnata* is a common succulent in the family Crassulaceae. Its leaves have a very bitter, or acidic, taste in the early morning, lose their sourness by the late afternoon, and regain it by the next morning. The biochemical steps underlying the results of Heyne's "taste test" are now referred to as *Crassulacean acid metabolism,* or *CAM* for short. Actually, CO_2 uptake at night by cladodes of a platyopuntia had been discovered in 1804 by Nicolas-Théodore de Saussure, a Swiss scientist whose broad interests included plant chemistry and physiology, but the details of the accumulation of organic acids in the chlorenchyma of agaves and cacti at night and their depletion during the daytime were not fully understood until the 1980s. The interrelationship between anatomy and physiology for water conservation by CAM plants is crucial for their ecological success and greatly enhances their agricultural promise in arid and semiarid lands.

REFERENCES

Benson, L. 1982. *The Cacti of the United States and Canada.* Stanford University Press, Stanford, Calif.

Felger, R. S., and M. B. Moser. 1985. *People of the Desert and Sea.* University of Arizona Press, Tucson.

Gentry, H. S. 1982. *Agaves of Continental North America.* University of Arizona Press, Tucson.

Gibson, A. C., and P. S. Nobel. 1986. *The Cactus Primer.* Harvard University Press, Cambridge, Mass.

Grant, V., and K. Grant. 1979. The pollination spectrum in the southwestern American cactus flora. *Plant Systematics and Evolution* 133:29–37.

Hubstenberger, J. F., P. W. Clayton, and G. C. Phillips. 1992. Micropropagation of cacti (Cactaceae). In *Biotechnology in Agriculture and Forestry.* Volume 20, *High-Tech and Micropropagation IV* (Y. P. S. Bajaj, ed.). Springer-Verlag, Berlin. Pp. 49–68.

Nobel, P. S. 1992. Annual variations in flowering percentage, seedling establishment, and ramet production for a desert perennial. *International Journal of Plant Sciences* 153:102–107.

Richards, H. M. 1915. *Acidity and Gas Exchange in Cacti.* Carnegie Institution of Washington, Washington, D.C.

Robberecht, R., and P. S. Nobel. 1983. A Fibonacci sequence in rib number for a barrel cactus. *Annals of Botany* 51:153–155.

Sánchez-Mejorada R., H. 1982. *Some Prehispanic Uses of Cacti Among the Indians of Mexico.* Gobierno del Estado del México, Toluca, Mexico.

Smith, J. P., Jr., ed. 1988. *Inventory of Rare and Endangered Vascular Plants of California,* 4th ed. California Native Plant Society, Berkeley.

2

Agaves: Food, Beverages, and Fiber

Because catching animals was difficult and often seasonal, early Native Americans used nearby plants as the staples of their diet, and in the process, they became extremely knowledgeable about agaves and cacti. About half of the 9000-year-old mummified human feces found in caves in Tamaulipas and Tehuacán, Mexico, contained remnants of both agaves and cacti, attesting to the consumption of such plants, and ancient artifacts of clothing and tools indicate many other uses for agaves. Whether prehistoric Americans cultivated these plants or simply collected them from nature is harder to determine. But the early encounters between humans and these remarkable plants set the stage for their many uses in the twentieth century, the focus of this chapter.

Roasting pits were used to cook many species of agave, including the common desert agave (*Agave deserti*), throughout the southwestern United States and northwestern Mexico (Figure 2.1). Native Americans competed with bighorn sheep and deer for the young inflorescences of *A. deserti*. Young flower buds of *Agave utahensis*, a species currently being considered for the U.S. threatened species list (Chapter 1), were roasted in earthen pits in present-day Arizona and Utah. *Agave palmeri* at lower elevations and *A. parryi* at higher elevations were harvested for roasting by Indians in Arizona. The Paiute Indians chewed the roasted leaf and stem morsels until only the fibers remained. Roasted inflorescences were pounded into "cakes of mescal" (*mescal* is another common name for certain agaves and also for a distilled alcoholic beverage made from them). After drying in the sun, these cakes were traded with local army posts in the nineteenth century. The sap exuded from tapped plants plus the chopped-up roasted leaves were fermented into an alcoholic beverage, and rope was made from the leaf fibers. Indeed, one Indian subtribe had so many uses for agaves that they became known as the Mescalero

Figure 2.1 Remnants of a roasting pit used by Yaqui Indians to roast *Agave deserti* in Anza Borrego State Park, California. (Photograph is courtesy of Mark Jorgensen.)

Apaches. The dried leaves of dead agave rosettes burn easily and have been used for household fires, signal fires, smoke signaling over longer distances, and fires associated with cattle herding. *Quids* (chewed pieces of a plant material that are not swallowed) from agave leaves have even been used as wads placed over gunpowder in muzzle loaders.

The spine at the tip of an agave leaf can be removed with an attached string of vascular tissue, creating a "needle and thread" combination. *Vascular tissue* generally occurs as tough fibers that strengthen a plant and act as conduits or pipes for the movement of water and nutrients obtained by roots from the soil and the movement of photosynthetic products originating in the shoots. Fibers from *maguey*, another common name for agaves from central Mexico, were used prehistorically for weaving and may have been among the earliest fibers so used in North America. In addition to remnants of agave leaves in mummified feces and quids, maguey fibers more than 5000 years old have been found in rope, bags, mats, baskets, sandals, clothes, and even paper. Certain magueys may have been some of the first cultivated plants in North America.

Food

The key to making agaves palatable is their high sugar content, and the key to breaking down the naturally occurring starchy tissues into sugars is heating. Roasting transforms complex *polysaccharides* (long chains of sugars, such as glucose, that form starch) into individual sugars that taste sweet and can be readily digested. Men and boys traditionally gathered the mescal heads, or *cabezas*, which refer to the stems with the leaves cut off but with the leaf bases still attached. Because a cabeza looks like a large pineapple, it is also called a *piña*, particularly for the agave used for tequila, *Agave tequilana*. Women and girls traditionally did the cooking in multilayered pits in a way similar to that described for the Seri Indians in Chapter 1.

Cabezas, young inflorescences, and even agave leaves for roasting or baking in pits are still harvested in northwestern Mexico and the southwestern United States, but less often today than at the turn of the century, when 30 species of agaves were harvested. Roasting cabezas is also a cottage industry that produces an enjoyable sweet treat somewhat like the stalks of sugarcane. In the desert countryside, an observant passerby can often detect rings of soot-covered stones surrounding the roasting pits used in the past (Figure 2.1). Roasted agaves are eaten like artichokes—that is, the leaves are scraped against the teeth to pull off the edible part, leaving the fibers behind—and the cabeza can be eaten in its entirety, like the heart of an artichoke.

In certain parts of Mexico, agave flowers are harvested for food. The tall inflorescences of species such as *Agave bovicornuta* and *A. stricta* are pulled over, and the side branches with their young flowers are removed. The flowers are often lightly cooked and then dipped into an egg batter for frying. Another delicacy from agaves is the grubs or borers that attack leaf bases and stems. The larvae of the brownish moths in the genera *Agathymus* and *Megathymus* are highly prized by the harvesters and are usually consumed fried. Agave borers, which can be 3 centimeters (1.25 inches) long, are even put into bottles of mescal exported from Mexico.

The cuticle and the underlying layer of epidermal cells can be peeled from the inner leaves of various species of agaves, such as *Agave atrovirens*, *A. mapisaga*, and *A. salmiana*, to form a translucent sheet in which food can be wrapped. This wrapping subtly flavors the delicacy *mixiote* (Plate C), which is prepared for special occasions, including Christmas and Easter. For mixiote, the agave cuticle is used as a wrap for steam-cooking packets of meat (especially chicken), chopped cactus stems (*nopalitos*), other vegetables, and spices. No agaves are raised specifically for this purpose, which generally leads to poaching of the cuticles. Indeed, the covert gathering of agave cuticles has risen to an art form in the Valley of Mexico, a fertile, relatively flat region of 7500 square kilometers (2900 square miles) extending from Mexico City to the northeast. Agaves cultivated there for beverage production can have leaves up to 2 meters (7 feet) long! The cuticle is easily removed from young leaves still folded about the central spike of unfolded leaves (Figure 2.2).

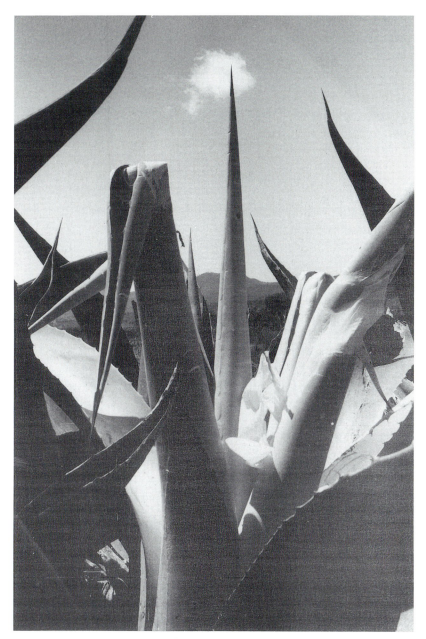

Figure 2.2   Agave salmiana, whose outer leaves bandits have pulled back from the central spike of folded leaves to remove the cuticle, used as a wrap for mixiote (Plate C). Note the piece of cuticle near the bottom right of the central spike of folded leaves. The plants were growing near Calpulalpan, Tlaxcala, Mexico.

Bandits surreptitiously move into the fields of cultivated agaves at night, pull the leaves away from the central spike of unfolded leaves, and cut off large rectangular sections of the cuticle from these separated leaves, often using the terminal spine of an agave leaf as the cutting tool (Figure 2.2). To prevent this devastation, farmers who raise agaves for pulque production often mutilate their own plants, rendering them unfit for cuticle harvesting. By lopping off the top one-third of the central spike of folded leaves with a machete, the farmer makes the cuticle too short to use for mixiote. In addition, holes made in the central spike with a pitchfork render the cuticle of unfolded leaves unsuitable for wrapping food, because the delicious juices would leak out during cooking. Such deliberate mutilation of the agaves discourages the rampant and seasonally lucrative thievery, but at a cost to plant productivity. The pitchfork treatment is less detrimental to productivity than is the removal of part of the central spike, as would be expected from the relative leaf area lost in the two cases.

Aguamiel and Pulque

The key to producing beverages from agaves is their accumulation of carbohydrates such as sugars and starch. Carbohydrates accumulate in the stems and the leaf bases—the best accumulators usually have the fattest stems and the thickest leaf bases. The oldest beverage still produced from agaves is *pulque,* which was known to the Aztecs (Figure 1.1) at the beginning of the thirteenth century and to earlier peoples more than a thousand years before that. There was even an Aztec goddess of pulque, Mayahuel, and pulque was used in religious rites that entailed human sacrifices. Pulque was commonly used by Aztec priests, who were given special dispensations. Specifically, people under 52 years of age were allowed to drink up to two cups a day, people over 52 could drink three cups a day, but priests could drink up to five cups a day, which often led them to a state of ecstasy! Yet drunkenness was not tolerated, with the third offense being punished by death.

Pulque is produced by fermenting *aguamiel* (literally, honey water), which accumulates in the hollowed-out stems of mature agaves, especially *Agave mapisaga* and *A. salmiana* (Figure 2.3). Aguamiel is a sweet beverage consumed mainly by women and children but generally enjoyed by all. It can rapidly turn to vinegar or be coaxed to ferment into pulque. Pulque production in Mexico ranges from an interesting and colorful cottage industry of single-household operations to large commercial haciendas, mainly in the nine states that surround Mexico City. The resulting pulque varies considerably from place to place. Pulque is usually drunk like beer or a hard apple-cider and contains a similar alcohol content (about 3 to 4 percent, with considerable variation). Pulque also contains some sugars, a wide range of amino acids, and many vitamins. To increase its market appeal, some pulque is flavored

Figure 2.3   Scraping the basin carved out of the center of a pulque-producing agave, *Agave mapisaga. Aguamiel* accumulating in the basin is removed twice daily for 2 to 6 months. The plant was growing near Calpulalpan, Tlaxcala, Mexico.

with fruit extracts, including banana, cherry, and pineapple. Pulque can also be used in making bread as well as distilled into a brandylike "aca pulque."

To produce pulque, a mature 10- to 12-year-old plant of species such as *Agave atrovirens, A. mapisaga,* and *A. salmiana* is selected from a plantation or from planted hedges around fields. Eight other species of agave are used to lesser extents, including species harvested from the wild and *Agave americana* (Figure 1.2). Agaves up to 20 to 25 years in age may be used, as long as the plants are large and nearly ready to flower. The central spike of folded leaves is usually entirely removed, leaving a hemispherical cavity at the center of the plant (Figure 2.3). Sometimes a large opening is cut into the base of the folded leaves, again creating a basal hemispherical cavity. Twice a day the liquid is removed, generally by sucking it into a gourd (Plate D). Each time liquid is removed from the cavity, the exposed basin is scraped to prevent the formation of calluses, which would slow the exudation of additional liquid. The sweet juice can be removed for 2 to 6 months, with one plant yielding from 400 to over 1000 liters of aguamiel. This pleasant-tasting liquid is still an important source of uncontaminated water in certain arid regions. Aguamiel is reputedly good for those with diabetes mellitus (it contains considerable amounts of the sugar fructose) and anemia, but it must be consumed within a few hours of collecting, as it sours quickly.

After being removed from the plant, the aguamiel is transported—usually by burro in two, approximately 30-liter, wooden casks strapped over its back (Plate D)—to a *tinacal* for brewing to make pulque. At tinacals, practices and quality control vary substantially. In the better ones, the aguamiel is mixed with a starter culture of old pulque for about 1 week, placed with more aguamiel into large tanks for another week or so, and then transferred into even larger vats for its final fermentation. This overall process requires more than 1 month. In lesser-quality tinacals, the starter stage takes only about 1 day, and the overall process lasts only a few days. In all cases, great care is taken to minimize contaminating the fermenting brew (Figure 2.4). The readiness of the final beverage, which becomes slightly carbonated by the fermentation process, is tested visually by evaluating the consistency of drops falling off a ladle dipped into the vat. A "synthetic" pulque can be made by diluting one part of real pulque with two or more parts of water, adding granulated cane sugar, and allowing the fermentation to continue for about 1 week.

Pulque is not bottled and must be drunk within a few days, or it will spoil. This whitish, viscous, slightly carbonated beverage is consumed mainly by working-class men. It is consumed and/or sold in *pulquerías* to which the carry-out buyer usually brings a container (Plate E). In 1993, approximately 30,000 hectares (70,000 acres) of agaves were cultivated in Mexico for pulque production, leading to about 400 million liters of pulque produced annually (1.00 liter = 1.06 quarts).

Mescal and Tequila

Distillation was introduced into Mexico in the sixteenth century from Europe and apparently also from the Philippines, resulting in two addi-

Figure 2.4   Vat of fermenting pulque near Calpulalpan, Tlaxcala, Mexico.

tional alcoholic beverages made from agaves, mescal and tequila. These beverages have rapidly gained worldwide economic importance. Indeed, agaves can no longer supply all the required starting material for tequila. Sugar from corn (maize) and sugarcane is sometimes fermented with the mashed cabezas, or alcohol (ethanol) produced from sugarcane is added to the final distillate, at least by certain distilleries. Both tequila production and mescal production begin with agave cabezas (Figure 2.5), but the technologies, species, and localities involved differ. In 1993, approximately 20,000 hectares (50,000 acres) of mescal agaves were cultivated (with some intercropping of corn and beans), leading to an annual mescal production of 20 million liters. In addition, 18,000 hectares of tequila agaves were cultivated, leading to 70 million liters of tequila produced annually.

Mescal has traditionally been produced in the Mexican state of Oaxaca, about 400 kilometers (250 miles) southeast of Mexico City. Mescal production is spreading, such as to the northwestern state of Sonora. The agave of choice is *Agave angustifolia*, but ten other species have been collected from the wild for making mescal, especially *Agave salmiana*. Although the planting of *A. angustifolia* for mescal production has increased, especially in Oaxaca, as many as 500,000 cabezas are obtained annually from the wild. Harvesting from the wild has potentially severe ecological consequences, especially in Sonora and other parts of northwestern Mexico, where *A. angustifolia* and *A. palmeri* are used for mescal.

Figure 2.5   Harvested cabezas (stem bases) of *Agave tequilana* awaiting roasting for the production of tequila. The plants were raised near the factory in Tequila, Jalisco, Mexico.

Determining which agaves are about to flower is essential to successful mescal production. The apical meristem hidden within the central spike of folded leaves is removed from such plants so that all subsequent energy derived by photosynthesis in the leaves leads to larger and larger cabezas. These plants, which cannot flower, may not be harvested for a year or more. Some plants are harvested without the prior excision of the apical meristem, a practice that is usually less efficient with respect to the accumulation of sugars and starch in the stem.

In some regions, the harvested cabezas are roasted in earthen pits for 3 to 5 days and then are crushed by a large millstone turned by a tractor or even a burro. The mashed cabezas are inoculated with a sample from a previous brew and allowed to ferment. The resulting liquid is distilled in a small still heated by a wood fire. The first liquid to condense from the still, the "head," can contain considerable amounts of the harmful methanol, and the last distillate, the "tail," has little of the desired ethanol and often contains various bad-tasting organic compounds. The purchaser can bring a container and obtain the mescal directly from the producer at a relatively low price. These cottage industries are being replaced by modern factories complete with carefully cultured inocula, assembly-line bottling practices, and recognized brand names. Fortunately, some of the old art has been retained, and even commercial mescals have a wide range of flavors and marketing strategies. Mescal can be doubly distilled, meaning that the original distillate (minus most of the head and the tail) is put through the still twice. The best mescal is doubly distilled and aged, usually in the bottle for up to about 4 years.

Tequila is doubly distilled and usually aged, in some cases for 4 or more years. Nearly all the factories are located in Tequila, Jalisco, about 500 kilometers (300 miles) west-northwest of Mexico City. The first tequila was manufactured in 1621 in Tequila, where the current technology was developed about 150 years ago. The agave of choice for tequila production is *Agave tequilana*, particularly its variety called *azul* (Spanish for "blue"), named for the bluish tint of its leaves. The cabezas (Figure 2.5) weigh 25 to 45 kilograms (55 to 100 pounds) each and are from plants approximately 8 years old. They are cooked for 36 to 48 hours in large ovens, which converts most of the accumulated starch to sugars. The cooked cabezas are mashed, shredded, and then fermented—generally in stainless-steel vats with carefully cultured yeasts for about 2 days—to convert the sugars to ethanol. The resulting fermented material is distilled in copper stills under strict quality control. The final product is usually diluted with water before being aged or bottled.

After the dilution during processing, tequila generally has an alcoholic content of 40 to 45 percent by volume (80 to 90 proof). The better grades of tequila are often consumed "neat" (straight up, without mixers). However, tequila is probably more famous as an ingredient in the cocktail called a *margarita*. The margarita was invented by Carlos Herrera in the 1950s in Tijuana, Baja California Norte, for Margarite, an American showgirl who did not like to drink tequila neat. Today, margaritas can be ordered in bars around the world. A margarita typically contains roughly two parts of tequila, one part of lemon or lime juice, and one-fifth to one-half part (depending on the sweetness desired) of the orange liqueur Triple Sec. The ingredients are blended with crushed ice and strained or poured directly into a glass whose rim has been moistened with a citrus rind and then dipped into salt. In 1992, about 4 million liters of tequila were exported from Mexico to the United States, and tequila is now exported to 70 other countries as well. To protect its tequila industry from foreign competition, Mexico has placed an embargo on the exportation of propagation stocks of *A. tequilana*.

Fiber

The strength of the agave fiber, which runs the entire length of the leaves, has long been known. Indeed, such fibers have been used for rope, twine, nets, sacks, furniture webbing, mats, cushion stuffing, upholstery padding, saddle pads, carpet pads, blankets, baskets, bracelets, headbands, sandals, clothing and other woven objects, armor, brushes, fish stringers, musical instruments, ceremonial objects, and, most recently, construction material, paper pulp, and dart boards! Paper also has been made from the fibers in the tender inner leaves of certain agaves. No wonder the utility of agaves was recognized in the early nineteenth century by European invaders in Mexico.

Europeans introduced the agave-fiber industry into their colonies, including Indonesia, the Philippines, and eastern Africa. In these regions, the main species used was sisal (*Agave sisalana*), which was also cultivated in Caribbean islands, many Pacific islands, Australia, and especially Brazil. *Agave sisalana* was never very popular for fiber production in its native Mexico, where henequen (*Agave fourcroydes*) was the main commercial fiber agave. At the turn of the nineteenth century, Mexico's main export income was derived from the leaf fibers of henequen (Figure 2.6). In the twentieth century, the fibers from agave leaves were used for agricultural products, such as baling twine for hay and sacks for cereal grains.

Mexico

The most widely harvested agave currently used for fiber in Mexico is lechuguilla (*Agave lechuguilla*). It is annually harvested or processed by more than 200,000 people and has more rosettes in the wild than does any other agave. Lechuguilla generally occurs on hillsides in nine states of east-central and northeastern Mexico as well as in Texas and New Mexico. It is the dominant perennial on about 16 million hectares (40 million acres). The preparation of the lechuguilla fiber is a cottage industry; nearly all the plants are harvested from the wild, especially in Coahuila, Nuevo León, San Luis Potosi, Tamaulipas, and Zacatecas. The fiber is obtained from mature unfolded leaves and the central spike of folded leaves, both of which, in mature plants, are about 60 centimeters (24 inches) long. To extract the fiber, the thorns on the leaf margins and

Figure 2.6   Plantation of 7-year-old plants of henequen (*Agave fourcroydes*). The lower leaves of the plants have been harvested and transported to decorticating plants for the removal of hard fiber in a factory near Mérida, Yucatán, Mexico.

the spine at the leaf tip are removed, the leaves are then pounded, and the pulp is scraped away with a knife. Finally, the fibers are placed in the sun to dry and to bleach the chlorophyll.

More widely distributed in nature than *A. lechuguilla* is *Agave angustifolia*, which can be found throughout Central America and most of Mexico. Leaves from plants in the wild and from plantations, such as in Guatemala, have also been used to produce fibers. In addition, commercial leaf fiber is obtained from *Agave rhodacantha* and *A. striata* harvested from the wild and from *A. sisalana*, which is now grown on a limited basis in eastern Mexico. Even though *A. lechuguilla* is more widely harvested and *A. angustifolia* has a wider distribution, the main commercial fiber in Mexico and the adjacent Caribbean islands comes from henequen (*Agave fourcroydes*; Figure 2.6). Henequen, whose origin is uncertain, is generally considered to be a sterile hybrid, although it sometimes produces fertile seeds. It is the dominant agricultural crop in Yucatán, the state at the tip of the eastern peninsula of Mexico, and is also grown in southern Tamaulipas and in Veracruz.

When henequen is around 5 or 6 years old, approximately 15 to 20 mature lower leaves are harvested annually per plant. The leaves, which are usually about 1 meter (39 inches) long, are cut off every few months and transported to large mills where the fiber is removed. In some mills, the leaves are centered on the stationary metal bar of a decorticator (a machine that removes some outer or less useful part of a plant). Two other bars descend and crush the leaf, leaving the long hard whitish fibers draped over the stationary bar while the leaf pulp drops beneath. The leaf juices can drip through a grating or can be mechanically squeezed out of the leaf pulp. They form an important by-product containing sapogenins (to be discussed shortly), and the pulp can be fed to cattle. Commercial-quality leaves are harvested for about 10 years, after which the production of an inflorescence, or "bolting," signals the end of the plant's life. Although henequen produces bulbils, its offshoots produced on rhizomes grow more rapidly. These offshoots must be cleared periodically from the harvest lanes between rows in the plantations, thereby simultaneously providing the propagation stock. In 1993, approximately 200,000 hectares (500,000 acres) of henequen were cultivated in Mexico.

Eastern Africa

The main agave raised for fiber in eastern Africa, as well as in Brazil, India, and various other Asian countries, is sisal (*Agave sisalana*), which is a sterile hybrid. In the first half of the twentieth century, sisal supplied about 70 percent of the world's long hard plant fibers—used for rope, twine, and "burlap" sacks—compared with 15 percent supplied by henequen and 15 percent by other species like abaca, also known as Manila hemp (*Musa textilis*). Since then, the sisal industry has declined worldwide because of the advent of synthetic fibers developed during and immediately after World War II. Synthetic fibers are uniform, strong, and resistant to water, whereas natural fibers generally become much weaker when wet and are prone to rotting. Yet in 1965, 910,000 tons of hard plant fiber were produced: 24 percent in Tanzania, 21 percent in other eastern African countries, and 20 percent in Brazil from sisal; and 16 percent in Mexico from henequen. Other agaves cultivated for fiber include *Agave cantala* and *A. lurida* in India and the Philippines. By 1985, the production of hard fibers from agaves was about half as much as in 1965, and today it is even less. Nevertheless, the history of making a plant from the New World a commercial success in the Old World is interesting in its own right.

Sisal was first exported from the port of Sisal in Yucatán, although the species apparently originated in the neighboring state of Chiapas. To protect its fiber industry, Mexico placed an embargo on the export of *Agave sisalana* in the middle of the nineteenth century, but some plants had already been exported and became established in Florida in 1836. In 1893, Richard Hindorf, an agronomist with the Deutsch–Ostafrikanische Gesellshaft (German East Africa Company), arranged to have 1000 bulbils of sisal sent from Florida to Germany. Although 200 bulbils survived that trip, only 62 survived the ensuing trip around South Africa to German East Africa (present-day Tanzania). Remarkably, after only 5 years, the 62 bulbils led to 63,000 plants, which radically changed the agricultural practices of eastern Africa. Sisal became an important crop in the countries presently known as Angola, Kenya, Madagascar, Mozambique, Tanzania, Uganda, and Zimbabwe.

Each sisal plant produces about 220 leaves before bolting. The leaves were often transported to central decorticating plants by means of small railway cars on narrow-gauge metal tracks. As they were for henequen in Mexico, mechanical decorticators were crucial for the success of sisal production in eastern Africa. Some decorticators are fed by hand. The pulp is first rasped from half of a leaf; the leaf is withdrawn; and then the opposite half is inserted for rasping. Those machines in which the leaves enter broadside are more efficient, as two raspers act simultaneously, or more usually in sequence, to remove the leaf pulp from the proximal (basal) and the distal halves of the leaves. Beautiful tresses of long strong fibers emerge from the decorticators. The fibers are usually washed and then dried for a few hours in the sun.

As do most agaves and cacti, sisal grows best on free-draining non-saline soils. Sisal bulbils tend to grow faster and more reliably than do their offshoots, which is the opposite case of henequen. Eastern Africa has the proper soils for agave cultivation, and vegetatively produced bulbils are readily available for sisal, but the sisal industry is currently in decline owing to the greater popularity of synthetic fibers. Changes in the availability of synthetic fibers and the development of secondary products will have a great effect on the future of agave-fiber plantations worldwide.

Other Uses

In addition to sources of food, beverages, and fiber, agaves have a long history of other uses. Many species of agave—such as *Agave atrovirens*, *A. ghiesbreghtii*, *A. hookeri*, *A. tecta*, and *A. weberi* as well as the common agaves *A. americana*, *A. mapisaga*, and *A. salmiana*—were planted as fences or hedges to separate fields. In some parts of Mexico, agaves, with their armor of terminal leaf spines and lateral leaf thorns, were planted as defenses against invaders. The leaves of agaves, including henequen (*A. fourcroydes*) and *Agave salmiana*, have recently served as food for dairy cattle, both penned and free ranging, and agave inflorescences are fed to range cattle in Baja California. The beautiful geometric symmetry of agaves has endeared them to plant collectors worldwide. The future economic importance of agaves may also be influenced by their high content of sapogenins, discussed later.

Ornamental Horticulture

Early Spanish and Portuguese visitors to the New World were enthralled by agaves, and the century plant *Agave americana* (Figure 1.2) soon made its way to the Azores, the Canary Islands, and the European continent. Since then, agaves—including the picturesque variegated form of *A. americana* that has longitudinal yellowish stripes alternating with dark green stripes along its leaves (Plate F)—have become favorite decorative plants in botanical and private gardens around the world. *Agave americana*, which is readily propagated vegetatively by ramets, has also been used as a fence or hedge on five continents.

Other agaves were also prized in the eighteenth and nineteenth centuries for their beauty: *Agave angustifolia*, *A. cantala*, *A. lurida*, and *A. salmiana* appeared in both public and private gardens in Europe, Asia, and Africa. The impressive *Agave filifera* has fibers emanating from its leaf margins (Plate G). Another popular agave is *Agave attenuata*, which has beautifully arching, long-but-broad, thin, grayish green leaves. Not only can it survive many months of drought, but its rather long stems produce numerous offshoots in response to rainfall or watering, making *A. attenuata* easy to propagate vegetatively. *Agave vilmoriniana*, also produced from bulbils, is increasingly used as an ornamental. Other agaves common in botanical gardens, private collections, and the horticultural trade include *Agave desmettiana*, *A. ellemeetiana*, *A. multifilifera*, *A. patonii*, *A. shawii*, *A. victoriae-reginae*, and the magnificently striped *A. zebra*.

Sapogenins

A wide range of special chemical compounds can be found in the leaves of agaves. Although leaf juices from many species were used topically by Indians for relieving itching and sores, the leaf juices of other agaves, such as *Agave bovicornuta*, cause dermatitis. Some agaves are still used as fish poisons, and an extract of lechuguilla (*A. lechuguilla*) was used to poison arrow tips. Such toxic compounds in the leaves of agaves may have evolved to repel both insects and larger herbivores. Juices pressed from the leaves of *Agave difformis*, *A. lechuguilla*, *A. toumeyana*, *A. vilmoriniana*, and especially *A. schottii* are still used as soaps. The active ingredients for these various effects are most likely sapogenins (except possibly in *A. bovicornuta*, whose skin irritant is unknown).

The *sapogenins* extracted from agave leaves are steroids (a complex group of organic molecules containing a 4-ring system with 17 carbon atoms) and are usually bound to a sugar, such as glucose. This sapogenin—sugar combination is called a *saponin*, which foams when shaken and is generally poisonous to certain lower animals, including fish. The saponin's foaming activity is related to its detergent property—the steroid part of the molecule is soluble in fats, and the sugar part is soluble in water. Hence, saponins can disrupt cell membranes, which contain fats, apparently leading to their lethal effect on certain animals. Sapogenins can also be used as a starting material in the manufacture of steroid drugs. The widely prescribed anti-inflammatory agent cortisone

(Figure 2.7A) as well as the chemically related female sex hormones estrogen and progesterone are steroids, with structures similar to those of agave sapogenins.

The structures of complex organic molecules bewilder nearly everyone (organic chemistry is not the most beloved of subjects). Yet we can readily see that there are numerous structural similarities between the common steroid cortisone (Figure 2.7A) and the sapogenin hecogenin (Figure 2.7B). Hecogenin occurs in the leaves of sisal (*Agave sisalana*) and many other agaves and can be recovered from the leaf juices of henequen (*Agave fourcroydes*) during fiber extraction. Sapogenins can make up more than 2 percent of the leaf dry weight for *Agave capensis, A. cerulata, A. nelsonii, A. promontorii, A. roseana,* and *A. vilmoriniana. Dry weight,* a term that we will use repeatedly, refers to the weight remaining after the water has been removed—for example, by heating the leaves in an oven at 80°C (176°F) for a few days. Dry weight is the preferred unit for many properties, including plant productivity. The total weight, generally referred to as the *fresh weight,* is measured while the plant is still alive or just after it has been harvested. The fresh weight varies with the plant's natural water content, which can change considerably with environmental conditions.

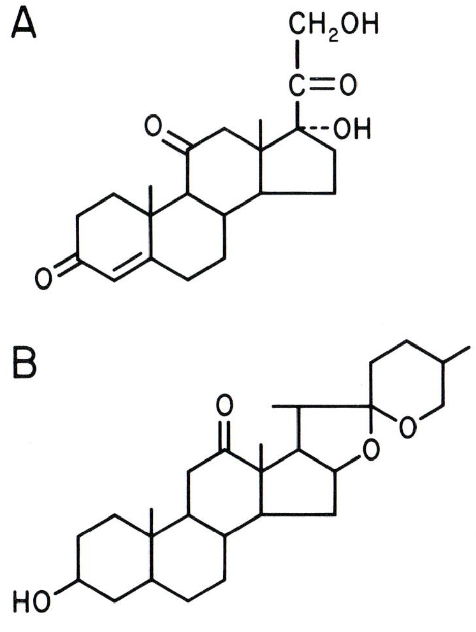

Figure 2.7 Chemical structures of steroids: (A) the anti-inflammatory drug cortisone; and (B) the sapogenin hecogenin, common in the leaves of many agaves.

The champion sapogenin producer is *Agave vilmoriniana*, whose leaves can contain up to 4.4 percent sapogenin by dry weight. When it reaches maturity and flowers in 8 to 10 years, *A. vilmoriniana* is a fairly large plant whose leaves can contain more than 3 percent of the sapogenin smilagenin. Smilagenin, which differs from hecogenin (Figure 2.7B) by only one bond (the double-bonded oxygen near the center is absent), has also been isolated from the leaves of lechuguilla (*Agave lechuguilla*). Steroids have a multitude of effects on both humans and other animals, so the potential uses of agave steroids in a crude form and after chemical modification are remarkably ripe for investigation. Already, 6 percent of the world's supply of precursors for corticosteroids comes from agaves. New applications for steroids can be expected to increase the commercial importance of agave sapogenins, and thus agave cultivation, in the future.

REFERENCES

Bahre, C. J., and D. E. Bradbury. 1980. Manufacture of mescal in Sonora, Mexico. *Economic Botany* 34:391–400.

Cruz, C., L. del Castillo, M. Robert, and R. N. Ondarza, eds. 1985. *Biología y aprovechamiento integral del henequén y otros agaves*. Centro de Investigación Científica de Yucatán, A.C. Mérida, Yucatán, Mexico.

Ebeling, W. 1986. *Handbook of Indian Foods and Fibers of Arid America*. University of California Press, Los Angeles.

García-Moya, E., and P. S. Nobel. 1990. Leaf unfolding rates and responses to cuticle damaging by pulque agaves in Mexico. *Desert Plants* 10:55–57.

Gentry, H. S. 1982. *Agaves of Continental North America*. University of Arizona Press, Tucson.

Lock, G. W. 1969. *Sisal: Thirty Years' Sisal Research in Tanzania*. Longmans, London.

Luna, Z. R. 1991. *La historia del Tequila, de sus regiones y sus hombres*. Consejo Nacional para la Cultura y las Artes, Mexico City.

Parson, J. R., and M. H. Parsons. 1990. *Maguey Utilization in Highland Central Mexico: An Archaeological Ethnography*. University of Michigan, Ann Arbor.

Pinkava, D. J., and H. S. Gentry, eds. 1985. *Desert Plants* 7:34–116 (special issue, Symposium on the Genus Agave).

Sheldon, S. 1980. Ethnobotany of *Agave lecheguilla* and *Yucca carnerosana* in Mexico's Zona Ixtlera. *Economic Botany* 34:376–390.

3

Cacti: Many Uses . . .
and a Few Mistakes

Cacti in general and their fruits in particular played many important roles in the daily lives of Native Americans (Table 1.1). The fruits were either eaten raw or after being cooked in earthen pits by a method similar to that used for roasting agave inflorescences (Chapter 2). The cooked fruits were stored or sometimes boiled, salted, and mixed with corn flour. Native Americans also dried cactus fruits and seeds in the sun; such products could last for a year or more and were readily transportable. A "cactus moon" was celebrated in the early spring, when the flower buds of opuntias like *Opuntia echinocarpa* and *O. versicolor* were cooked or roasted as a special treat. Harvesting fruits from saguaro (*Carnegiea gigantea*; Figure 1.7B) and other cacti often became a race among humans, rodents, birds, and insects. An Indian in the Sonoran Desert could harvest about 350 kilograms (770 pounds) of saguaro fruit per season. Indeed, this fruit harvest was so important to the Papago and Pima Indians that harvest time (June) was the beginning of their calendar year. Wine produced from the saguaro fruit was also enjoyed by Indians in what is now the southwestern United States. The saguaro has even been championed as "by far the most noteworthy representative of plant life in the desert, being, in fact, one of the most remarkable plants on the globe."*

In addition to eating the fruits and flowers of cacti, Native Americans also ate their stems. Young cladodes, termed nopalitos, were cooked as a vegetable and often combined with eggs and the meat of wild animals. Barrel cacti (also called *biznagas*) like *Echinocactus platyacanthus* and *Ferocactus histrix* provided water, material for candy, and containers for food and were even used as tables for human sacrifices. The spines of various

*Carl Lumhotz, *New Trails in Mexico* (New York: Scribner's, 1912).

45

species were used as fishhooks by many and sometimes as toothpicks by the Aztecs.

Fruits

Native Americans seem to have tasted the fruit of nearly every species of cactus, as is extensively documented in the ethnobotanical literature. The fruits of many cacti are delicious and are usually eaten raw. Local cottage industries have developed throughout Latin America to sell fruits collected in the wild from more than 40 species of cacti in 15 genera. In time and with improved marketing strategies, some of these species may assume greater economic importance and may become commercially produced. Cactus fruits are currently experiencing a rise in popularity in the United States. Over 10,000 tons were imported from Mexico in 1992, including a red variety of prickly pear (Plate H) in addition to the more common fruits that, when ripe, are yellowish green both inside and outside. Cactus fruits are high in sugars (generally 70 to 80 percent by dry weight). About one-third of the sugar content is fructose, which is better tolerated than glucose and sucrose by those with diabetes mellitus. Cactus fruits are also high in vitamin C (ascorbic acid) and low in fats.

Prickly pear fruits are often called "cactus pears" and "pear apples" in the United States, *tunas* in Latin America, "Barberry figs" and "Indian figs" (*ficus indica* in Latin) in many European countries, and *sabras* (connoting thorny outside but sweet inside) in Arabic and Hebrew in northern Africa and southwestern Asia. They are also called "pest pears" in Australia for reasons that will be made clear later. Because the name "prickly pear" has negative connotations, the name "cactus pear" is gaining in popularity for the marketing of this fruit. We will henceforth use "cactus pear" when referring to the fruit, reserving "prickly pear" for the plant.

Before eating cactus pears, the glochids are usually first brushed off. These nasty small spines can be removed by rolling the fruits on the ground, swatting the fruits with small branches of leaves, or wiping them with a moist towel. Glochids are removed commercially by using mechanical brushes. Indeed, retention of glochids is a major factor that limits the acceptance of cactus fruits in new markets. To facilitate peeling, a small amount of the top and the bottom of each fruit is cut off, and a slit is made from end to end through the rind, or peel. Next the peel is pulled back, exposing the juicy, often brightly colored, delicious pulp. Or the whole fruit can simply be cut in half and the pulp scooped out with a spoon. If one does not mind swallowing the seeds—and they are innocuous when ingested—eating the pulp is easy.

The seeds can also be a food source, even when collected from human feces (sometimes called a "second harvest"). The seeds are eaten after being roasted, or they are dried in the sun and then ground into a flourlike meal used for cooking. Seeds from *Carnegiea gigantea, Ferocactus covillei, F. wislizenii,* various *Opuntia* species, *Pachycereus pectenaboriginum, P. pringlei, Stenocereus gummosus,* and *S. thurberi* are consumed in the Sonoran Desert of northwestern Mexico and the southwestern United States. Cactus seeds also are used in animal feed, usually in a ground form. In addition, oil for cooking and even for industrial purposes has been extracted from the seeds of 30 species of *Opuntia*.

Cactus Pears

Christopher Columbus may have returned to Spain with *Opuntia ficusindica*. Subsequent Spanish conquerors definitely took prickly pear cacti back to Europe as an ornamental curiosity and later as the host for the cochineal insect (discussed in this chapter). Originally planted in the roof gardens and orchards of the aristocracy, prickly pears soon became naturalized across southern Europe. They spread near cultivated fields by vegetative means as well as over larger distances by birds that fed on the fruits. When the Moors were forced out of Spain in 1610, they took prickly pear cacti with them to northern Africa. The cacti eventually spread eastward to Israel (Figure 1.7C). Prickly pears are still an important fruit crop in arid and semi-arid regions bordering the Mediterranean Sea, ranking in popularity with grapes and olives.

In Mexico, the main species cultivated for fruits are *Opuntia amyclea, O. ficus-indica* (Plate H), *O. joconostle, O. megacantha,* and *O. streptacantha;* species collected from the wild for their fruits include *O. hyptiacantha, O. leucotricha,* and *O. streptacantha.* Indians also ate the fruits of these species and, to a lesser extent, their young cladodes. For easier harvesting, they collected desirable cactus species from remote areas and planted them closer to their homes, which led to hybridization between species normally separated in nature, thereby creating new cultivars. The taxonomy of some highly desirable cultivars has thus become hopelessly tangled. Also because of cultivation, the native habitat of certain species has become unclear. For example, today we do not know the place of origin of the most widely grown prickly pear, *Opuntia ficus-indica*.

In the Northern Hemisphere, cactus pears usually mature from July through October. The seasonal offset of approximately 6 months in the Southern Hemisphere leads to fruit harvests in January through April in countries like Chile and South Africa. The delayed fruit harvest developed in Sicily (discussed shortly) can supply cactus pears for the remaining months, and so cactus pears can be available year-round. The fruits of the main cultivated species, *O. ficus-indica,* are pleasantly sweet and are generally liked by first-time tasters. The edible pulp makes up 60 to

75 percent of the fruit and usually contains 12 to 15 percent sugars by fresh weight, within the generally acceptable range for commercial fruits eaten raw.

In addition to cactus pears eaten raw, cactus fruit is prepared in many other ways in Mexico, including stewing. The seeds often are removed by pressing the peeled fruit through a colander. The resulting pulp is boiled and thus reduced in volume, creating a thick syrup. When cooled, this syrup becomes a popular taffy called *miel de tuna* (literally, cactus pear honey). Miel de tuna is widely sold in the markets, sometimes as the syrup. The pitted juicy pulp can also be vigorously stirred, boiled for about 8 hours, cooled, and then allowed to form cakes called *queso de tuna* (literally, cactus pear cheese). Queso de tuna is made primarily from the fruits of *Opuntia streptacantha* and is sold in a brown, bricklike form. The freshly peeled whole fruits can be hung on a string, and after a couple of weeks in the sun, these pulps become coated with a sweet sticky juice and can last for many months without further preservation. The pulps and especially the peels of the fruit of *Opuntia joconostle* are used in many ways, including in cooking and candy making. *Melcocha* is a thick jelly made by reducing the volume of the fruit juice by boiling. A fermented drink, *colonche* (or *nochote*), is also prepared from the juice of the fruit. Wine from the fruits of *O. robusta* and *O. streptacantha* is distilled to produce a brandy. Removing young fruit (and cladodes) often causes a gummy sap to exude from the cut surface of the remaining cladodes. This sap can be eaten raw, roasted, or mixed with water as a beverage, as is often done with sap from the chain-fruit cholla (*Opuntia fulgida*). For many species, the seeds, which are high in fats, are consumed after being roasted or ground.

In 1992, approximately 60,000 hectares (150,000 acres) were cultivated worldwide for the production of cactus pears (Table 3.1). Over 80 percent of these lands are in Mexico, which also has the greatest diversity of cultivated species. Chile utilizes prime agricultural land near Santiago for its well-developed cactus pear industry, whose fruits are widely available in Chilean supermarkets. Cactus pear orchards were planted in Bolivia in the 1980s, and small plots of prickly pears can be found in Argentina. In South Africa, nearly as much fruit is sold in roadside stands and by street vendors as in the commercial markets. In Israel, the informal market of cactus pears harvested from hedges (Figure 1.7C) and small plantings is also about the size of the commercial market (Table 3.1). Small plantations in the United States about 100 kilometers (62 miles) southeast of San Francisco, California, supply high-quality, individually waxed and wrapped cactus pears.

Two crops of cactus pears often can be harvested annually if the soil is kept moist by rainfall plus supplemental irrigation. The annual fruit harvest is typically 5 to 10 tons fresh weight per hectare (2 to 4 tons per acre). Harvests can be 15 tons per hectare on five continents—North

Table 3.1 Representative Land Areas Under Cultivation for Fruit or Fodder Production by *Opuntia ficus-indica* and Similar Species

Country	*Fruit* Area (hectares)*	Annual harvest (tons fresh weight)
Argentina	500	2,500
Bolivia	1,200	3,000
Chile	1,100	8,000
Israel	300	6,000
Italy	2,500	50,000
Mexico	52,000	300,000
South Africa	—	300
United States	120	—

	Fodder (hectares)
Brazil	>300,000
Mexico	230,000
South Africa	30,000
Tunisia	70,000

Note: Data are for 1992 and were obtained from many sources.

*One hectare equals 2.47 acres.

America, South America, Europe, Africa, and Asia—and 22 tons per hectare when two major harvests are produced per year. Even higher harvests are possible under careful management. Annual fruit yields of at least 30 tons fresh weight per hectare have been achieved in Chile, Israel, and South Africa.

The Sicilian Experience

We cannot leave our discussion of cactus pears without describing a practice that began in Sicily by accident—the removal of flowers in early summer—and that had remarkable consequences. As one legend goes, the advances of a neighbor's son toward a farmer's daughter so enraged the farmer that he knocked off all the flowers from his neighbor's prickly pears. Another legend has a farmer from a village near Palermo refusing to sell his cactus pears to a neighbor, who in revenge also removed all the flowers. A third legend has a son removing some flowers from the prickly pear cladodes to increase the size of the subsequent fruits, and in response, the angry father removed the rest of the flowers.

Whatever its origin, flower removal, somewhat surprisingly, causes the cacti to flower again. Although fewer flowers are produced the second time, they develop into larger and sweeter fruits that are harvested in the late autumn. These late fruits also have less market competition from fruits of other species, leading to higher prices. The importance of this late market has institutionalized the practice of *scozzolatura*, which in Italian means "to take the buds away" (Figure 3.1). In June, laborers now move through Sicilian fields of cultivated *Opuntia ficus-indica* with ladders and sticks, deliberately destroying the first fruit crop! The plants flower again in 30 to 40 days, and the next fruit crop ripens from the end of October into early December. The edible tissue of such fruits varies in color: most (80 to 90 percent) are yellow (*gialla*), 8 to 15 percent are red (*rossa*); and 2 to 5 percent are white (*bianca*). Fruits are also graded by size: Extra large are over 160 grams fresh weight (5.6 ounces); first class are 120 to 160 grams (4.2 to 5.6 ounces); second class are 80 to 120 grams (2.8 to 4.2 ounces); and third class, which have low market appeal, are under 80 grams.

Figure 3.1   Removal of flowers from *Opuntia ficus-indica* in the early summer, a practice called scozzolatura, which leads to a more valuable fruit crop harvested in the late autumn. This commercial orchard is near Santa Margherita Belice, Sicily.

In 1992, Sicily had about 2500 hectares of cactus pear orchards on which scozzolatura was practiced (Table 3.1). These regions have sandy soils and a Mediterranean climate—that is, mild rainy winters and hot dry summers. The annual rainfall of 500 to 600 millimeters (20 to 24 inches) is sometimes supplemented in the summer with irrigation equivalent to about 70 millimeters of rainfall. The spacing of the cacti is similar to that for small fruit trees, at intervals of 4 to 5 meters (13 to 15 feet) along rows that are 5 to 7 meters (16 to 23 feet) apart. With the practice of scozzolatura (Figure 3.1) and sometimes subsequent thinning, 8 to 10 fruits mature per terminal cladode of *O. ficus-indica*. The fruits are harvested by gloved hand or with scissors and plastic cups to catch the fruit. Cactus pears are also harvested informally in Sicily without scozzolatura on about 25,000 hectares of hillsides, roadsides, and gardens. Nearly 3000 tons of fruit are exported annually from Sicily, mainly to Belgium, France, Canada, and the United States. Because of Sicily, Italy is the main exporter of cactus fruits in Europe.

Other Species

Although not as well known as the fruits from prickly pear cacti, the fruits from columnar cacti and barrel cacti are also collected for human consumption. These fruits are mainly eaten raw, but they are also used to prepare the equivalents of miel de tuna, queso de tuna, melcocha, colonche, and wine, as for cactus pears. The fruits and the columnar cacti themselves are generally referred to as *pitahayas* or, more commonly, *pitayas*. The most important genus of pitayas for fruit harvesting in Mexico is *Stenocereus* (Tribe Pachycereeae). Seven species of *Stenocereus* are cultivated for their fruits, and in 1993 these included approximately 500 hectares (1200 acres) of *S. griseus* planted in Puebla, Oaxaca, Tamaulipas, and Veracruz and 1000 hectares of *S. queretaroensis* in Jalisco, Michoacan, and Queretaro. These pitayas have a fruit yield of 5 to 10 tons per hectare; over 80 percent are red inside and outside; and most are harvested in May. The fruits from another three species of *Stenocereus* are collected from the wild, and the fruits of saguaro (*Carnegiea gigantea*; Figure 1.7B) have long been important in northwestern Mexico and southwestern United States. The fruits of seven other species of tribe Pachycereeae are harvested from the wild, including *Pachycereus pecten-aboriginum*, *P. pringlei*, and especially *Escontria chiotilla*, whose fruits are used for marmalade and ice cream. Another Pachycereeae, *Myrtillocactus geometrizans*, is increasingly planted in central Mexico for its fruits.

Although less popular, the fruits of species in other tribes of subfamily Cactoideae are also harvested in Mexico, including *Hylocereus undatus* in tribe Hylocereeae, a few species of *Echinocereus* in tribe Echinocereeae (for example, the strawberry cactus [*E. stramineus*]), and various species of *Ferocactus* and *Mammillaria* in tribe Cacteae. Floral buds of various barrel cacti are also consumed, often fried with eggs or boiled with chilis. Either the fruits of *Ferocactus histrix*, commonly sold in Mexican markets, are eaten raw or their juice is extracted for use in popsicles, marmalades and syrups, and even wine.

In the 1980s, commercial plantations occupying about 1200 hectares (3000 acres) were developed in Colombia for the extremely delicious fruits of *Hylocereus triangularis* (also known as *H. megalanthus* and sometimes *H. undatus*). Five years after planting, yields reached 6 tons of pitayas per hectare for the two annual harvests. The fruits (Figure 3.2) are usually yellow outside and white inside, weigh about 300 grams (0.7 pound), are hand-brushed to remove the spines, and have a thin peel. Field trials in Israel have selected various fruit varieties of *H. triangularis*, including a fruit that is red outside with white pulp and varieties that are red or yellow both outside and inside. Such fruits can be both tasty and large, often 600 grams. Selling fruits of *Hylocereus costariensis* and *H. polyrhizus* is a cottage industry in Central America and northern South America, with great potential for future large-scale commercialization. *Hylocereus* is also raised in Vietnam, and its fruits are transported to city markets in China. Another cactus whose fruits have great market potential is *Cereus peruvianus*, which is widely cultivated as an ornamental. It produces large edible fruits (up to 400 grams each) that vary from red to yellow outside with white pulp, soft seeds, and a delicious flavor.

Figure 3.2   Fruits of pitaya (*Hylocereus triangularis*) in a supermarket in Bogotá, Colombia.

Young Cladodes as Vegetables

Young stem segments (cladodes) of prickly pear cacti are used as a green vegetable throughout Mexico and in the southwestern United States. The whole cladodes are commonly called "joints," "pads," or "nopales" (sometimes they are erroneously referred to as leaves). Once prepared for eating, the sliced or diced cladodes are almost always called "nopalitos." Various species are used for nopalitos, primarily *Opuntia ficus-indica, O. robusta,* and *O. streptacantha,* plus *O. rastrera* in the Chihuahuan Desert. Sometimes species of *Nopalea,* including *Nopalea auberi, N. cochenillifera,* and *N. karwinskiana,* are also used. Milpa Alta, located just southeast of Mexico City, is the major region for growing *O. ficus-indica* and other platyopuntias for the important nopalito market of Mexico City (Figure 3.3).

The cladodes are often sliced into strips about 6 millimeters (0.25 inch) wide. They are then cooked with onions, peppers, cheese, eggs, and spices to make delicious fillings for tacos and other dishes. The strips can also be marinated before being cooked. Bottled marinated nopalitos are now commonly sold in grocery stores throughout the southwestern United States and increasingly in other locations. Marinated nopalitos can be eaten directly as an hors d'oeuvre, placed in salads, or used in cooking. Cladodes can also be diced into cubes about 1 centimeter (0.38 inch) on a side and then put into omelettes or used as a vegetable in other dishes, such as the mouth-watering treat mixiote (Plate C).

The key to preparing delicious nopalitos is the starting material. Tender young cladodes a few weeks old are picked when they are thin, dark green, and 15 to 30 centimeters (6 to 12 inches) long. The spines at this stage are often inconspicuous, and the glochids are not yet the menace that they are for mature cladodes. Moreover, the areoles, on which the spines and glochids arise, are prominently raised above the young stem. With a few deft strokes using a sharp knife that is moved at a grazing angle to the cladode surface, the glochids and other parts of the areole can thus be removed without the loss of much stem tissue. This is frequently done in front of the nopales purchaser in Mexican markets (Figure 3.3). A potato peeler can also be used to remove the areoles, which, although slower, requires less skill.

A cladode is usually sliced or diced, depending on the desired texture and cooking time. The stems can be simmered until tender (about 10 minutes) in roughly one-third their volume of water with salt and sometimes onions, garlic, and cilantro. If any sticky material remains after draining, the nopalitos can be rinsed with cold water. There are over 200 recipes for these vegetables, whose taste is pleasant and whose texture is appealing. Some say the taste is like gherkins or green peppers, and the texture is between that of string beans and okra. Accep-

Figure 3.3   Street market where glochids are removed from the young cladodes of *Opuntia ficus-indica* and the *nopalitos* are sold as a vegetable. The cladodes were harvested near the market in Milpa Alta on the outskirts of Mexico City. The woman is using a knife to remove glochids from the cladode margin.

tance among first-time users is high, and veteran users may eat nopalitos seven days a week.

The nopalito market in the United States is confined mainly to the Southwest, where approximately 400 tons were consumed annually in the 1980s in Arizona, California, and Texas. Sales are particularly strong during the springtime season of Lent (the 46 days from Ash Wednesday to Easter). Nearly all nopales are imported from Mexico and are eaten by people of Mexican descent, although the popularity of nopalitos is spreading. The annual commercial production of nopalitos in Mexico in 1992 was about 250,000 tons fresh weight grown on 6000 hectares (15,000 acres). Under careful management, with harvests every 15 days, more than 400 tons fresh weight hectare^{-1} year^{-1} have been produced in Milpa Alta (Figure 3.3).

Forage and Fodder

Various species of cacti have been used as fodder, including large barrel cacti in the genus *Ferocactus* and some columnar cacti, although platyopuntias are the most commonly harvested. *Fodder* refers to coarse plant material that is cut, transported, and then fed to livestock such as cattle, horses, sheep, goats, and pigs. For barrel cacti used as fodder, the stem is often simply cut open, exposing the pulp, whereas the stems of columnar cacti, such as *Neobuxbaumia tetetzo* and *Pachycereus pringlei*, are generally chopped. The ecological consequences of harvesting naturally occurring cacti can be devastating, because species in genera like *Ferocactus* tend to propagate infrequently and grow slowly. Accordingly, large regions in Baja California Norte and certain islands in the Golfo de California, Mexico, have been denuded of barrel cacti to provide emergency rations for livestock. Luckily for preservation, most of the cacti used for livestock fodder are cultivated platyopuntias, such as *Opuntia ficus-indica* in central Mexico and *O. engelmannii* and other species in northern Mexico and southern Texas (Figure 3.4).

In Mexico, range cattle, deer, and other animals also eat cylindropuntias as forage, including *Opuntia bigelovii*, *O. fulgida*, and *O. versicolor*, as well as platyopuntias like *O. phaeacantha*. *Forage* refers to plant material eaten by animals where the plants grow. Opuntias can be the principal component of the annual diet of white-tailed deer, representing 21 percent of their food intake. Quail and wild turkey eat opuntias less frequently, but javelinas, various rodents, and some turtles and tortoises depend more heavily on opuntias as a food source. Barrel cacti are also a crucial source of water for desert animals, including pack rats, gophers, rabbits, and bighorn sheep. Especially in dry years, sheep paw and butt over various species of *Ferocactus* in the southwestern United States and northwestern Mexico in a relentless pursuit for water.

From about 1901 to 1915, the flamboyant American plant breeder

Figure 3.4   Santa Gertrudis bull eating *Opuntia engelmannii* on the Maltsberger Ranch, Cotulla, Texas. In 1940, Santa Gertrudis became the first recognized breed of cattle to be developed in the United States. (Photograph is courtesy of William A. Maltsberger.)

Luther Burbank promoted the fruits of various prickly pears for human consumption and their cladodes for cattle fodder. The prices for his products were sometimes rather high, a few dollars for a single cladode of his spineless variety. His claims were also often exaggerated, such as an annual fruit production of 500 tons fresh weight per hectare (200 tons per acre). Burbank wrote that the development of prickly pear cacti "promises to be of as great or even greater value to the human race than the discovery of steam."* Although various species of opuntia can be essentially spineless and earlier breeders had developed spineless platyopuntias with promising productivity, Burbank did help develop an important spineless variety of *Opuntia ficus-indica*. However, its virtues as a cattle fodder were overstated by promoters. Newly planted fields of this spineless prickly pear were common in the 1920s near Los Angeles, California, but feeding cladodes to cattle was minimal, in part because an overall program had not been planned. The same euphoria over the potential of prickly pear cacti for fodder extended to Australia and South Africa in the early twentieth century, with grave consequences that we will soon consider.

The importance of cacti as a source of both food and water has long been known to cattle ranchers in the southwestern United States. The

Los Angeles Examiner, 19 July 1911.

use of prickly pear cacti as cattle feed in southern Texas and northern Mexico has been documented since 1857. More recently, ranchers in southern Texas have planted platyopuntias on a limited basis, mainly as a water source for their cattle during recurring droughts. For ranchers to plant cacti is almost treason, as ranchers are not farmers, and the local prickly pear cacti still are usually considered to be weeds. The practice of removing the spines by singeing the cacti with a "pear burner" (Plate J), a modified blow torch, was begun toward the end of the nineteenth century. Pear burners today use propane typically carried in a 5-gallon (19-liter) tank on a backpack. One gallon of propane can burn the spines off enough cladodes to feed four or five cows for a day, not an insignificant feat considering that cows can consume each day up to 10 percent of their body weight in prickly pear cladodes. Cattle readily eat the singed cladodes, which supply not only sugars needed for energy, but also water. Cattle in Sonora, Mexico, and in southern Texas can be sustained for many months of drought without any additional water, relying instead on the water in the prickly pear cladodes.

The water and nutritional contents of prickly pear cladodes depend on environmental conditions. After a few months of drought, the water in the cladodes of wild platyopuntias may be only 60 percent of the cladode fresh weight, whereas *O. ficus-indica* irrigated daily can contain 96 percent water by fresh weight. A typical water content for cladodes in the field is 87 to 90 percent by fresh weight. Nitrogen obtained from the soil is required for the synthesis of chlorophyll, proteins, nucleic acids, and many other molecules necessary for the functioning of plant cells. The nitrogen content of stems of the cacti used for forage and fodder ranges from about 0.7 to 2.5 percent of the dry weight. The higher range of nitrogen in cacti is the same as the nitrogen levels in the best forage grasses and herbs, although most platyopuntias used for forage have 0.8 to 1.4 percent nitrogen, which corresponds to 5 to 9 percent protein, by dry weight. The protein percentage needed to sustain cattle usually ranges from 6 to 10 percent (the higher levels are required for lactating cows). Hence, cladodes fed to cattle as their primary food source are customarily supplemented with a protein-rich material such as cotton seed or bone meal. A nutritionally promising variety of *Opuntia stricta* has been identified that can have nitrogen and phosphorus levels that are fully adequate for cattle, and nutrient levels in cacti can be raised by appropriately fertilizing the plants.

Phosphorus in the cladodes of forage platyopuntias generally averages 0.1 to 0.3 percent of the dry weight. This phosphorus content is slightly below the average nutritional needs of cattle, especially those of lactating cows. The phosphorus nutrition of cattle is also affected by the high levels of calcium and oxalate in cacti, which reduce the availability of phosphorus for metabolism. Sodium can be below 0.01 percent of cladode dry weight, whereas cattle require about 0.1 percent sodium in their diets. Both phosphorus and sodium are therefore supplemented for livestock consuming large amounts of prickly pear cacti. Soluble sugars and other carbohydrates needed for energy constitute about 50 percent of cladode dry weight, a somewhat higher percentage than is present in hay or alfalfa. The fiber content of the cladodes is relatively high, averaging 30 percent by dry weight, which is sufficient for range cattle and also beneficial in human diets.

Various species of prickly pear cacti are raised worldwide for fodder, especially the spineless *Opuntia ficus-indica* (Table 3.1). Large areas are planted for fodder in Mexico. Spineless *O. ficus-indica* is cultivated in South Africa, mainly as fodder for cattle and sheep during drought. Large plantations also can be found in Tunisia and other countries in northern Africa. The largest areas devoted to prickly pear for fodder are in northeastern Brazil, mostly for *Opuntia cochenillifera* and *O. ficus-indica* in the states of Alagoas, Paraiba, and Pernambuco. In this region, with its short rainy season followed by a drought lasting about 8 months, prickly pear can be a major source of water for livestock during most of the year.

Royal Red: The Cochineal Story

The robes of Aztec emperors, including Montezuma, were a deep royal red. To create this color, the emperors demanded that their subjects pay a tax in insects—specifically, the insects containing the vivid red dye (Plate I). In the sixteenth century, the colorful Aztec robes caught the fancy of Spanish conquerors, and platyopuntias were soon shipped to the shores of the Mediterranean and, in 1831, to the Canary Islands for raising the cochineal insect. At one point, the dye was worth more than its weight in gold!

The insects used for commercial production of the red dye feed on the cladodes of certain prickly pears, including *Opuntia ficus-indica* and *O. tomentosa*. The dye comes only from the female cochineal insects in genus *Dactylopius*, which surround themselves with telltale cottonlike webs on the cladodes (Figure 3.5). These females, which are about 3 millimeters (0.125 inch) long and 2 millimeters across, are much larger than the males, which develop into winged adults. The females, however, remain in a prolonged larval stage and can draw nutrients from the cladodes for up to 3 years via their tubular mouthparts.

The source of the dye was a closely guarded secret of the Spanish traders who introduced the dye to Europe. But in 1704, Anton van Leeuwenhoek, the Dutch scientist who refined microscopes and first clearly described cells, helped dispel the mystery when he noticed the unmistakable parts of an insect in the cochineal stain he was using. In its heyday, the dye was used to produce the vibrant red of the robes of European royalty as well as the striking red jackets of the British Regu-

Figure 3.5   The platyopuntia *Opuntia ficus-indica* with the cottony webs of the many cochineal insects (*Dactylopius opuntiae*) on the cladodes and the fruit in Los Angeles, California.

lars, referred to by the American patriot Paul Revere when he warned in 1775 that "the redcoats are coming." Red jackets dyed with the extract from the cochineal insect were also worn by the Northwest Mounted Police, or "Mounties" (now the Royal Canadian Mounted Police). In the eighteenth century, the value of the cochineal dye exported from Mexico was second only to that of silver. But in 1856, cheap synthetic aniline dyes were developed from coal tar, which revolutionized the dyeing industry and stole the market from the cochineal dye. Nevertheless, the stain carmine (or carminic acid) is still prepared from cochineal insects and is still very expensive.

To produce the dye, cochineal insects can be incubated on individual detached cladodes of *O. ficus-indica*. The cladodes are maintained in sheds to avoid the problems associated with rainfall and high winds. The insects can be removed by jets of air and collected mechanically before extracting the dye. However, most cochineal insects are currently harvested by hand from *O. ficus-indica* growing in the field. Up to 20 grams (0.7 ounce) of insects can be obtained from a heavily infested cladode (Figure 3.5). Cochineal insects are raised for carmine production in the Mexican state of Oaxaca, as well as in Algiers, the Canary Islands, Chile, Morocco, South Africa, and especially Peru. Carminic acid from cochineal insects is used as a natural coloring in foods, soft drinks, and many cosmetics (including lipstick); as a pH indicator (a visual detector of the acidity or alkalinity of a solution); and as a pigment prized by artists (Crimson Lake). Because coal-tar (aniline) dyes have been linked to cancer in laboratory animals, there is now renewed interest in the cochineal dye as a coloring for foodstuffs, including shrimp, jams, Maraschino cherries, and the Italian aperitif Compari. The annual worldwide commercial production in 1992 was 300 tons, about 90 percent of which came from Peru and the remainder primarily from the Canary Islands.

The cochineal dye is potent but costly to extract. The preferred dye is collected before the females lay their eggs (oviposit). Approximately 130,000 adult females must then be harvested to obtain 1 kilogram (2.20 pounds) dry weight of *grana negra* (black grains, the word *grain* owing to the Spanish mistaking the small insect bodies for seeds). Before ovipositing, about 80,000 females are required per kilogram of *grana blanca* (white grains). Imagine the time required to handpick so many insects from the cocoon webs on the cladodes, as the Aztecs did! The webs with the insects are harvested and dried in the sun. Sometimes the insects are killed in boiling water, which also dissolves their wax coating, before they are dried. The dried insects are then ground to produce a purplish powder, the most commonly marketed form. The dye, which constitutes 10 to 26 percent of the insects' dry weight, may also be extracted in boiling water and marketed in liquid form. Dried leaves may be added to the boiling water to loosely absorb the dye, after which they are dried and pressed into cakes for transport to traditional markets.

Peyote and Other Hallucinogens

Peyote (*Lophophora williamsii*) is a small, essentially spineless cactus about 8 centimeters (3 inches) in diameter that grows in northern Mexico and southern Texas. The word *peyote* comes from the Aztec name for this plant, *peyotl*. Only the top of its hemispherical stem is above ground. The stem of peyote can be cut into about four sections and then dried to make "buttons." The ingestion of such buttons, or of dried peyote flowers, leads to a state of mental exhilaration caused by the alkaloid mescaline (Figure 3.6A; *alkaloid* refers to a diverse group of nitrogen-containing organic molecules). Such effects, accompanied by colorful and often bizarre hallucinations, inspired the ritual use of peyote. Indeed, "peyotism" is a widespread practice directly influencing more than 200,000 Native Americans and others from Mexico to Canada. Peyotism blends tenets of Christianity with various Indian traditions and is practiced in the Native American Church. A federal permit is now required for possession of the peyote cactus, which can be used legally in religious ceremonies in most of the United States. When a large cactus collection of more than 200 species was recently offered to a university botanical garden, the collection was gratefully accepted—except for peyote, because of unsolvable security problems.

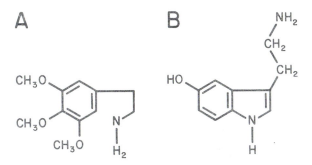

Figure 3.6 Chemical structures of two alkaloids: (A) mescaline, a hallucinogenic compound found in peyote (*Lophophora williamsii*) and various other cacti; and (B) serotonin, a compound that affects the transmission of nervous impulses in vertebrate animals.

Mescaline is chemically similar to compounds that affect the transmission of nervous impulses, such as the alkaloid serotonin (Figure 3.6B), a neurotransmitter that affects brain functioning in vertebrates. Like serotonin, mescaline stimulates the sympathetic nervous system, increasing blood pressure, dilating the pupils, and leading to arousal patterns in the brain. The effects of mescaline are similar to those of lysergic acid diethylamide (LSD), but mescaline is not as potent. Like LSD, mescaline often leads to periods of depression, headaches, and insomnia after the "high" wears off.

Hallucinogens are present in other cacti besides peyote (*L. williamsii*). The stems of *Opuntia cylindrica* from Ecuador and Peru can contain 0.9 percent mescaline by dry weight, and the fruits of *Opuntia leptocaulis* can be slightly hallucinogenic. The stems of saguaro (*Carnegiea gigantea*) also contain alkaloids similar to those in peyote. Other cacti that contain alkaloids specifically used by Indians for their chemical effects— especially to increase the endurance and visual perception of long-distance runners—include *Ariocarpus fissuratus*, *Coryphantha compacta*, *Echinocereus triglochidiatus*, *Epithelantha micromeris*, *Mammillaria heyderi*, and *Trichocereus peruvianus*. In Ecuador and Peru, San Pedro cactus (*Trichocereus pachanoi*), which reputedly has more mescaline than peyote does, has long been used medicinally and as a hallucinogen and can be purchased in street markets.

Other Uses

Cacti have an incredibly large number of uses, as demonstrated by the Seri Indians (Table 1.1). Many species of columnar cacti, including *Eulychnia acida* in South America (Figure 3.7) and *Pachycereus marginatus* in North America, are used for fences. Constructing a fence with the stems of columnar cacti is relatively easy. The stems are harvested

Figure 3.7   Eulychnia acida used as a fence near Barrancas, Chile.
(Photograph is courtesy of Herman Silva R.)

from the wild or from an established fence. The cut surfaces are allowed to dry and callus for a week or more. A stem is then inserted into a hole in the ground, and new roots grow in a matter of weeks. Platyopuntias, including *Opuntia stricta* in India and *O. ficus-indica* worldwide, are also widely used as fences or hedges. The wood remaining after the soft tissues have disintegrated from columnar cacti can be used for canes, and that from chollas such as *Opuntia fulgida* can be made into musical instruments and novelty decorations, including lamps and furniture.

The leaves of certain leafy cacti in the genera *Austrocylindropuntia* and *Pereskiopsis* can be eaten, as can the flower buds and flowers from many cactus species in various genera, including *Echinocereus, Ferocactus, Myr-tillocactus,* and *Opuntia.* Perfume is made from the flowers of various species of *Opuntia,* and coloring agents can be extracted from the fruits, such as those of *Opuntia phaeacantha.* Barrel cacti are a source of water, although drinking the juice directly pressed out of the stems can cause an upset stomach or diarrhea. A more palatable liquid can be obtained by evaporating the water from a mashed stem, condensing it on the walls of a container, and then allowing the condensed water to trickle into a collection vessel. As it becomes more widely known that certain species of cacti, especially platyopuntias, can be highly productive, the current uses may be expanded and many additional uses may become possible in the future, such as for cattle fodder, hillside stabilization, production of biogas (mainly methane) and gasohol, and even absorption of the increasing levels of atmospheric carbon dioxide (Chapter 8). We next consider other uses of cacti, as candy, ornamental plants, and the source of certain chemicals.

Candy

Cactus "candy" has many origins. Seri Indians used barrel cacti, especially *Ferocactus wislizenii,* as containers for honey collected from the wild (Table 1.1). After the honey was removed, children delighted in eating the sweet, honey-soaked walls of the stem. Another early cactus candy used sugars from saguaro fruit boiled with the water-storage parenchyma of the devil's pincushion cactus (*Echinocactus polycephalus*).

Many barrel cacti collected from the wild in Mexico, including *Echinocactus platyacanthus* and *Ferocactus histrix,* are still used to make candy. To make cactus candy, the water-storage tissue from large mature plants is cut into pieces, boiled in water for 2 to 3 hours, and cooled. Sugar is then added (both white and brown sugar are used), and the mixture is boiled again, this time for 1 to 2 days. When the pieces are cooled and allowed to air dry, the incorporated sugar, which weighs as much as the fresh cactus tissue, readily crystallizes. Cactus candy is eaten as is or is used in pastries, cakes, and even sweet tamales. Cactus candy is made in small-scale family operations using a single barrel cactus and in commercial enterprises using tons of plants at a time. Unfortunately, collecting cacti for candy has made *Ferocactus histrix* rare near towns and cities. Habitat deterioration by overgrazing has further threatened the existence of such barrel cacti.

Ornamentals

Perhaps no other plant family has as many representatives growing in homes throughout the world as does the Cactaceae, especially in regions with cold winters. The attractions of these cacti are many. The patterns of their spines and areoles are mesmerizing; they are slow growing and hence do not require frequent repotting; they are very forgiving of neglect, as watering can often be skipped for months at a time; and even small plants can produce gorgeous large flowers (Plates K and L).

About 300 species of cacti are now available as ornamentals. Small cacti in the following genera are extremely popular: *Ariocarpus, Echinocereus, Epiphyllum* (Plate L), *Gymnocalycium, Lobivia, Notocactus, Rebutia, Sulcorebutia,* and especially *Mammillaria.* The technology for propagating ornamental cacti is highly developed in Belgium, England, France, Germany, Italy, Japan, the Netherlands, Spain, and the United States, with large nurseries devoted to growing many varieties indoors from seed (Figure 3.8). The growth rate of cacti under optimal conditions of water and nutrients is generally much greater than under the often harsh conditions of their native habitats. Within 10 months of germination, stems can be 5 centimeters (2 inches) tall, a size that is readily accepted for sale in the retail trade. Some species destined for the ornamental markets are still harvested from the wild in Mexico and other Latin American countries, particularly species in the genera *Cor-*

Figure 3.8   Propagation of *Mammillaria supertexta* (left foreground),
Echinocereus stramineus (right foreground), and other species grown from seed
at C & J Nurseries in Vista, California. Most of the plants pictured are 6 to 12
months old, about 5 centimeters (2 inches) tall, and destined for the
ornamental cactus market.

yphantha and *Mammillaria*, although regulations of CITES (Chapter 1)
seek to avoid the potentially severe ecological consequences of this collecting. Members of cactus and succulent societies meet regularly to
discuss the protection of native species, as well as the discovery of new
species, propagation techniques, and improved methods for growing
cacti.

Medicine and Hormones

Extracts of various cacti have a wide range of medical applications (for
cacti utilized for medicinal purposes by the Seri Indians in northwestern
Mexico, see Table 1.1). Especially important today are the juices extracted from crushed young stems of *Opuntia* and *Lophocereus*, which are
used as antidiabetic agents in many parts of the world. The peels of the
fruit of *Opuntia joconostle* are also so used in Mexico. The juices from
young cladodes of *Opuntia stricta* and other platyopuntias are diluted
with water and then used as a diuretic in Australia, South Africa, and
many Latin American countries. *Mucilage*, a complex polysaccharide
composed of many uncommon sugars (such as arabinose and xylose), is
isolated from platyopuntias and *Pachycereus hollianus*. It has laxative
properties and also is used as a food thickener and an adhesive. Mucilage from the stems of *Opuntia imbricata* and *O. tunicata* is used to

increase the viscosity of pulque (Chapter 2). The cut stems of many species of cacti are used topically to treat burns and skin sores, and extracts of stems are used to cure stomach ulcers and kidney disease. Some medicinal claims for cacti, however, such as their ability to cure rheumatism and cancer, are exaggerated. Modern research techniques need to be applied to determine which chemical constituents of cacti are beneficial to human health, using experiences from traditional and current herbal medicine as a guide.

Both cacti and agaves contain steroids (Figure 2.7) that can act as hormones for humans and other animals. Such steroids occur in saguaro (*Carnegiea gigantea*; Figure 1.7B), senita (*Lophocereus schottii*), sina (*Stenocereus alamosensis*), organ pipe (*Stenocereus thurberi*), and other columnar cacti. There is an interesting relationship between various species of fruit fly (*Drosophila*) and the rotting stems of such cacti on which they feed. These cacti contain a compound used by the insects as a precursor to the steroid hormone required for molting (α-ecdysone). By feeding on the rotting stems of senita, *Drosophila pachea* obtains the steroid schottenol, which is chemically similar to α-ecdysone. The larvae of this fruit fly then require fewer chemical steps to make the hormone necessary for their development into adult flies. Senita also contains an alkaloid that is poisonous to other species of fruit fly. In turn, other columnar cacti contain alkaloids that are poisonous to *D. pachea*. Such relationships between a cactus species and a species of fruit fly have evolved over millions of years.

When Cacti Take Over

Contrary to the reputation of cacti as slow growers, platyopuntias accidentally or deliberately introduced into various regions have sometimes grown out of control. Certain features of these cacti encourage their spread at the expense of the local flora and despite the efforts of humans. A specific cactus not desired in a particular location is sometimes simply chopped up. But even small pieces of platyopuntias can develop roots and lead to new plants. Vegetative propagation can also occur when cladodes become detached by the wind or a passing animal. When the cacti produce fruit with fertile seeds, birds that eat the fruit can spread the seeds in their feces over great distances. Their spines can prevent grazing animals from eating cacti, and so spiny forms of platyopuntias spread relatively unchecked, outcompeting local flora that lack such a defense.

Australia

Beginning in 1832, platyopuntias were used as hedges for vineyards in the Hunter Valley, located 125 kilometers (80 miles) northwest of Sydney, Australia, and a single plant of *Opuntia stricta* was introduced into the Sydney area as a garden ornamental in 1839. Once introduced, the cacti became naturalized and could not be removed by being plowed under, which instead of reducing the number of cacti only created more. In 1883, legislation was passed to control *O. stricta*, which had by then become a dreaded pest (Figure 3.9). To make matters worse, in 1914, tons of cladodes of Burbank's spineless prickly pear cactus were shipped to Australia for forage. The plants grew well, flowered, and produced viable seed. The seeds, however, produced both spiny and spineless plants, characteristic of the ancestral breeding stock. Cattle and sheep would not eat the spiny plants, which therefore began to take over the land (Figure 3.9). By 1925, prickly pear cacti in eastern Australia, including *O. stricta*, *O. ficus-indica*, and *O. vulgaris*, were infesting new rangeland at the rate of 100 hectares (250 acres) per hour! Approximately 10 million hectares, mainly in Queensland, were infested.

To stop the spread of such cacti, the Australian government took action. In 1901, a reward of $10,000 was offered for a method to destroy the prickly pear cacti. The reward was doubled in 1907 but withdrawn in 1909 because of political bickering. The chemical poison arsenic was tried, with limited success. Beginning in 1912, government scouts were dispatched to Latin America to find the natural insect enemies of prickly pear cacti for use as biological control agents. Most of these enemies turned out to be various species of cochineal insects. Meanwhile, large

Figure 3.9 Infestation of *Opuntia stricta* near Chinchilla, Queensland, Australia, in the 1920s. (Photograph is courtesy of C. Barry Osmond, from an original held by the Queensland Department of Lands.)

nets and other barriers were erected to prevent the movement of birds that could spread the seeds in their feces. Innumerable hunting parties were sent out to shoot crows, emus, and scrub magpies. All together, 270,000 emus were killed. Such dubious means of controlling the spread of prickly pear proved ineffective and costly in many ways.

Most of the first biological control procedures were also ineffective against platyopuntias in Australia. The larvae of various species of cochineal insects did feed on the cladodes but did not greatly reduce the number of cacti. Larvae of the moth *Cactoblastis cactorum* were therefore introduced in 1925 and became the major control agent for prickly pear cacti in Australia. Larvae are produced twice annually from the approximately 400 eggs laid by a fertilized female. They can consume most of the chlorenchyma and part of the water-storage parenchyma of four mature cladodes before becoming moths themselves. By 1930, the moth was propagating naturally, and by 1933 fully 90 percent of the prickly pear plants had been eradicated by moth predation. But the near takeover of eastern Australia by the platyopuntias indicates their competitive ability and their high capacity for biomass production.

South Africa

The spineless form of *Opuntia ficus-indica* credited to Burbank was also brought into South Africa in the early twentieth century (1914) as potential cattle fodder. The spineless form again gave rise to its spiny ancestral version, as in Australia. Spiny forms of *O. ficus-indica* had been previously introduced as ornamentals and for hedges. The platyopuntia *Opuntia aurantiaca*, a hybrid that originated in eastern South America, had also been introduced earlier and had become a widespread weed in South Africa. To combat the spread of cacti, eradication efforts were undertaken using biological controls based on experiences in Australia.

At the peak of its spread in the early twentieth century, *Opuntia ficus-indica* infested 900,000 hectares (2.2 million acres) in eastern South Africa. Biological control in the 1930s included a species of the cochineal insect, *Dactylopius opuntiae*, which greatly reduced the prickly pear cactus in 75 percent of the infested area, but its multiplication was hampered by high rainfall and, to a lesser extent, by predators. The larvae of the moth *Cactoblastis cactorum* also killed a substantial number of small, isolated prickly pear plants, although it was much less effective in South Africa than on the relatively small plants of *Opuntia stricta* in Australia. This was in part because of egg loss through adverse weather conditions and greater predation on its larvae in South Africa. Minor additional control of *O. ficus-indica* was achieved with a weevil and another beetle.

Ironically, *O. ficus-indica* is now a valuable crop in South Africa, and farmers complain about the continuing effectiveness of the biological control agents introduced when it was considered to be only a weed. The spineless *O. ficus-indica* is currently raised on 30,000 hectares, primarily for fodder used during drought as well as for its fruits (Table 3.1). In an attempt to reverse the successful effects of biological control agents on prickly pear, control agents are now being directed against the control agents *D. opuntiae* and *C. cactorum*!

The many-jointed cactus (*Opuntia aurantiaca*), which is native to South America, has also been managed by biocontrol agents in South Africa. It apparently was introduced as an ornamental plant from England in the 1840s. By 1990, *O. aurantiaca*, whose joints are easily detached by animals, had infested nearly 900,000 hectares, mainly in the Eastern Cape and the Karoo, similar to the location and the total area infested by *O. ficus-indica*. *Dactylopius austrinus*, another cochineal insect species, was introduced as a biocontrol for *O. aurantiaca* in 1935 and soon became the officially recommended agent for eradication. Chemical control also became widespread in 1957, and both types of control are now used. Cochineal insects can be killed during heavy rainfalls, so the primary biocontrol agent is periodically ineffective. *Cactoblastis cactorum*, which was so effective on *Opuntia stricta* in Australia, was released in South Africa in 1933 to control *O. ficus-indica* and soon began attacking *O. aurantiaca* as well. Other species of moths have been released to control *O. aurantiaca*, but with limited success. Limited tests using fungal pathogens for biological control of cacti have been conducted in South Africa, and biocontrol has been attempted for a dozen other species of cacti that were introduced into South Africa as ornamental plants. Various species of cochineal insects, although attacking only a few species of these less common cacti, are proving to be the most effective biocontrol.

A North American Experience

Opuntia littoralis, *O. oricola*, and their hybrid (sometimes called *O. occidentalis*) are found on Santa Cruz Island, a 39-kilometer-long (24-mile-long) island situated about 40 kilometers off the coast of Santa Barbara, California. Overgrazing by introduced sheep facilitated the spread of these platyopuntias, whose cladodes rooted easily on the denuded ground (Figure 3.10A). About 40 percent of the island was eventually rendered unsuitable for grazing, which was the principal industry of the landowners. Accordingly, beginning in 1940, insects that infected prickly pear cacti in other parts of United States were brought to Santa Cruz Island to be tested as biological control agents for the platyopuntias.

Chelinidea tabulata and *C. vittiger* are beetles whose larvae and adults feed on the cladode sap, often leading to structural weakening and collapse of the plants. Their release on Santa Cruz Island, however, caused little damage to the prickly pear cacti. Three other species of

Figure 3.10   Before (A) and after (B) the release of the cochineal insect (*Dactylopius opuntiae*) among the platyopuntia *Opuntia oricola* growing on Santa Cruz Island, California. The insects were released in 1963. The second picture was taken in 1971 (see Goeden and Ricker, 1980). (Photographs are courtesy of Richard D. Goeden and the Commonwealth Scientific and Industrial Research Organization of Australia.)

insects were subsequently released but did not become established. Eventually, several species of cochineal insects (*Dactylopius* spp.) became established, leading to massive eradications of the platyopuntias in the 1960s (Figure 3.10B). One of the most effective species was *Dactylopius opuntiae*, which is native to southern California and northern Mexico. It was collected from Mexico in the 1920s and taken to Australia for biocontrol of prickly pear cactus. In 1949, its descendants were taken to Hawaii to control *Opuntia megacantha*. Cladodes infected with *D. opuntiae* were subsequently transported to Riverside, California, where the insects were cultured and then taken to Santa Cruz Island in 1951. This cochineal insect native to one part of southern California thus took a roundabout trip across the globe to help eradicate platyopuntias in another part of southern California only 240 kilometers (150 miles) away!

An important agent for spreading the insect-infected pads throughout Santa Cruz Island proved to be the sheep. By moving infected cladodes attached to their fleeces, they spread the cochineal insects from an infected patch to an uninfected patch of prickly pear. The cochineal insects multiplied without their natural enemies, which were hampered from reaching Santa Cruz Island by 40 kilometers of ocean. In interesting turns of fate, the denuding of the land and the moving of the cladodes by sheep initially favored the spread of the cactus; the moving of the infected cladodes by sheep spread the cochineal insects and led to the eradication of the cactus; and now sheep have been eliminated by humans in favor of cattle.

Native and introduced platyopuntias also occur in many other rangeland areas of the continental United States. For instance, prickly pear cacti grow on about 12 million hectares (30 million acres) in Texas. When overgrazing by cattle, goats, or sheep reduces annual and perennial grasses, these cacti tend to multiply and can become serious pests. Because mechanical eradication tends to spread the cacti vegetatively, the usual means of control have been grass fires and applications of herbicides, including aerial spraying of picloram. Fire can reduce the number of cladodes of prickly pear cacti by about 70 percent, but the plants begin to recover after 3 to 4 years. Fire plus low dosages of picloram (0.3 kilogram per hectare, 0.25 pound per acre) are more effective, leading to 98 percent eradication. The fire destroys most of the above-ground grass, allowing the herbicide to reach the ground. The springtime rains then move the picloram to the shallow roots of the platyopuntias *Opuntia engelmannii*, *O. lindheimeri*, and *O. phaeacantha* (sometimes considered to be the same as *O. engelmannii*). Ironically, at other locations in Texas, ranchers are planting prickly pear cacti as reserve forage to be used during times of drought, after the spines are burned off (Plate J).

The purpose of rangeland management is to encourage desirable vegetation and to discourage "uneconomic" species. Prickly pear is a natural part of many ecosystems, providing food and refuge for many species of wild animals. It is a major forage for deer, and fees to hunt deer are a substantial source of income on many ranches in South Texas. Efforts to control or to promote such cacti thus represent a continuing battle among conservationists, environmentalists, hunters, ranchers, and politicians.

REFERENCES

Barbera, G., F. Carimi, and P. Inglese. 1992. Past and present role of the Indian-fig prickly pear (*Opuntia ficus-indica* (L.) Miller, Cactaceae) in the agriculture of Sicily. *Economic Botany* 46:10–20.

Barker, J. S. F., and W. T. Starmer, eds. 1992. *Ecological Genetics and Evolution: The Cactus–Yeast–Drosophila Model System*. Academic Press, San Diego.

Benson, L. 1982. *The Cacti of the United States and Canada*. Stanford University Press, Stanford, Calif.

Bravo-Hollis, H., and H. Sánchez-Mejorada R. 1991. *Las cactáceas de México.*, Volume III. Universidad Nacional Autónoma de México, Mexico City.

del Castillo, R. F., and S. Trujillo. 1991. Ethnobotany of *Ferocactus histrix* and *Echinocactus platyacanthus* (Cactaceae) in the semiarid central Mexico: Past, present and future. *Economic Botany* 45:495–502.

Ebeling, W. 1986. *Handbook of Indian Foods and Fibers of Arid America*. University of California Press, Los Angeles.

Felker, P., ed. 1990. *Proceedings of the First Annual Texas Prickly Pear Council*. Texas A&I University, Kingsville.

Goeden, R. D., C. A. Fleschner, and D. W. Ricker. 1967. Biological control of prickly pear cacti on Santa Cruz Island, California. *Hilgardia* 38:579–606.

Goeden, R. D., and D. W. Ricker. 1980. Santa Cruz Island—revisited. Sequential photography records the causation, rates of progress, and lasting benefits of successful biological weed control. In *Proceedings of the V International Symposium on Biological Control of Weeds*. CSIRO, Melbourne, Australia. Pp. 355–365.

Granados S., D., and A. D. Casteñeda P. 1991. *El nopal: Historia, fisiología, genética e importancia fructícola*. Trillas, Mexico City.

Gregory, R. A., and P. Felker. 1992. Crude protein and phosphorus contents of eight contrasting *Opuntia* forage clones. *Journal of Arid Environments* 22:323–331.

Hanselka, C. W., and J. C. Paschal, eds. 1989. *Developing Prickly Pear as a Forage, Fruit and Vegetable Resource*. Texas A&I University, Kingsville.

Moran, V. C., and H. G. Zimmermann. 1991. Biological control of jointed cactus, *Opuntia aurantiaca* (Cactaceae), in South Africa. *Agriculture, Ecosystems and Environment* 37:5–27.

Osmond, C. B., and J. Monro. 1981. Prickly pear. In *Man and Plants in Australia* (J. Carr and S. M. Carr, eds). Academic Press, Sydney. Pp. 194–222.

Pimienta B., E. 1990. *El nopal tunero*. Universidad de Guadalajara, Guadalajara, Jalisco, Mexico.

Rangelander: Pricklypear Control Newsletter. Fall 1986. Dow Chemical, Dallas, Texas.

Ross, G. N. 1986. The bug in the rug. *Natural History* 3:67–73.

Russell, C. E., and P. Felker. 1987. The prickly-pears (*Opuntia* spp., Cactaceae): A source of human and animal food in semiarid regions. *Economic Botany* 41:433–445.

Tull, D. 1987. *A Practical Guide to Edible & Useful Plants*. Texas Monthly Press, Austin.

Uribe, J. M., M. T. Varnero, and C. Benavides. 1992. Biomasa de tuna (*Opuntia ficus-indica* L. Mill) como acelerador de la digestión aneróbica de guano de bovino. *Simiente* 62:14–18.

Wessels, A. B. 1988. *Spineless Prickly Pears*. Perskor, Johannesburg.

Zimmermann, H. G., and V. C. Moran. 1991. Biological control of prickly pear, *Opuntia ficus-indica* (Cactaceae), in South Africa. *Agriculture, Ecosystems and Environment* 37:29–35.

4

Roots: Water Uptake

Water is crucial for life. For a plant to grow, even for a seed to germinate, it must take up water, usually from the soil. The uptake of water and the minimization of its loss are two of the most important physiological processes for plants native to regions where water is scarce. We thus begin our consideration of the physiology of agaves and cacti with their roots, the site of water uptake.

The roots of agaves and cacti tend to grow at shallow depths in porous sandy soils. The light rainfalls that generally characterize arid and semiarid regions usually do not wet the soil very deeply, and thus shallow roots are ideally situated to respond quickly to the light rains. This, coupled with the water-conservation properties of the shoots of agaves and cacti, helps maintain a high water content in their shoots. Yet shallow roots can present a dilemma when the soil dries. The roots can then become conduits for massive losses of water from the shoots, which have a high water content, to a dry soil. In this chapter, we will consider various properties of roots and soil that facilitate water uptake into a plant under wet conditions but greatly limit water loss from the roots to a dry soil during drought. In engineering terms, the ready movement of something in one direction but not in the opposite direction is referred to as *rectification*. We will discover that the roots plus the soil act as a rectifier for water movement.

There are various types and shapes of roots, which reflect their different functions. Main roots emanating from the stem tend to be long, exploring a large volume of soil for both water and nutrients. Main roots can also help anchor a plant in place. Finer lateral roots branch from the main roots, thereby greatly increasing the area of contact between the root system and the soil particles. This large area facilitates the further uptake of water and nutrients from the soil. Sometimes a large fleshy root grows directly under the stem. These *taproots* (*tap-* comes from *tappe,* Middle English for a short tapering pipe or peg) can store large amounts of water and carbohydrates.

Root Distribution, Morphology, and Anatomy

The most striking characteristic of the root systems of agaves and cacti is their shallow depth, which has been carefully documented for desert succulents since the early twentieth century. For instance, William Cannon noted that the saguaro (*Carnegiea gigantea;* Figure 1.7B), which can exceed 10 meters (33 feet) in height and can live for 200 years, has roots with a mean depth of only 25 centimeters (10 inches)! He also excavated roots of *Ferocactus wislizenii* and *Opuntia discata.* Their roots can extend a few meters from the base of the plant but are always close to the soil surface, with a mean depth of 11 centimeters (4 inches).

Similarly, very few roots of *Agave deserti* (Figure 1.5) and *Ferocactus acanthodes* (Figure 1.7A) are deeper than 25 centimeters. The mean depth of their roots is only about 9 centimeters for *A. deserti* and *F. acanthodes* (Figure 4.1) and 8 centimeters for *Echinocereus engelmannii,* species that often occur together in the Sonoran Desert. The roots of these species, although shallow, are virtually absent from the upper 3 centimeters of the soil (except directly under the shoots). This upper region of the soil can become extremely hot during the summer. The

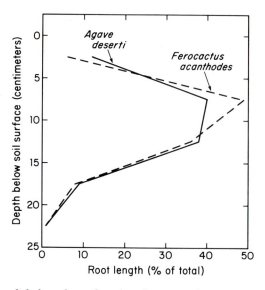

Figure 4.1 Depth below the soil surface for roots of two species from the Sonoran Desert, *Agave deserti* and *Ferocactus acanthodes.* Data are for 5-centimeter (2-inch) layers of soil at the site in Figure 1.5. For details, see Jordan and Nobel (1984) and Hunt and Nobel (1987).

surface of the soil where these species grow can reach 70°C (158°F), and the soil at a depth of 3 centimeters can reach 55°C (131°F). Such high temperatures generally prove lethal to root cells.

Although root depth is approximately the same for agaves and cacti, the morphology of their roots differs considerably. The monocotyledonous agaves are characterized by relatively straight roots that originate from the base of the stem. These roots tend to be thin, generally less than 4 millimeters (0.2 inch) in diameter, and have few, even thinner lateral roots. As an agave grows older, new roots develop from the base of its stem, but the existing roots do not become thicker with age. On the other hand, roots of the dicotyledonous cacti tend to branch and rebranch, and the main roots can become much thicker with age. The roots of 100-year-old saguaros can be more than 2.5 centimeters (1 inch) in diameter.

Various columnar cacti, including saguaro, have one or more main roots that penetrate vertically to deeper layers of the soil, in addition to their shallow, branching roots. These deep roots help to anchor the plant, and also they obtain water and nutrients. Many smaller cacti have a massive taproot extending vertically beneath the stem. For instance, *Copiapoa cinerea* (Figure 1.6) has a single large taproot in addition to a system of shallow roots extending horizontally in all directions. The fleshy taproot of peyote (*Lophophora williamsii*) has relatively few lateral roots branching from it. Various species of *Neoporteria, Peniocereus*, and *Pterocactus* have large tuberous roots that are many times larger than their stems. Such tuberous roots can store large amounts of water and carbohydrates underground. About 10 percent (160 species) of the Cactaceae are epiphytes that grow nonparasitically on other plants, chiefly tropical trees. The roots of epiphytic cacti can originate from many locations along the stem and tend to be rather thin, short, and highly branched.

For both agaves and cacti, wetting of the soil by rainfall induces the growth of new roots. Some of these roots emerge from the base of the stem, and so the number of main roots increases with the plant's age. For *A. deserti* growing in the northwestern Sonoran Desert, a 10-year-old plant produced on a rhizome might have about 20 main roots averaging 40 centimeters (16 inches) in length; a mature, 50-year-old plant typically has about 90 main roots averaging 80 centimeters in length. For the much faster growing *Agave mapisaga* in the Valley of Mexico (Figure 2.3), a 3-year-old plant has about 30 main roots averaging 50 centimeters in length, and a mature, 16-year-old plant has about 250 main roots averaging 170 centimeters (5.5 feet) in length.

Although new main roots continually form on agaves in wet soil, an isolated rainfall more commonly induces new lateral branches on the existing main roots. These branches persist as long as the soil is wet, but they die and fall off (*abscise*) when drought occurs. Rainfall also induces new lateral roots for cacti. However, the abscission of the small lateral branches for cacti is less common during drought than it is for agaves. The persistence of lateral roots of cacti, coupled with the general tendency for cactus roots to branch, leads to many fine roots. Only 6 months after a cladode of *Opuntia ficus-indica* (Figure 1.7C) is planted in wet fertile soil, over 80 percent of its root system dry weight can be from the relatively thin lateral roots.

The water taken up by a plant moves into and across roots through various cell types and by two different pathways. The *apoplastic* pathway, which involves water movement within the cell walls, begins with water entering the cell walls of epidermal cells (Figure 4.2). *Cell walls* surround and enclose all plant cells, are strong yet porous, and are composed mainly of tough fibers made up of a polymer called *cellulose*. The region between the cellulose polymers is filled with water. These water-filled regions allow water to move from the cell wall of one cell to the cell wall of an adjacent cell and thereby to cross the root cortex in the apoplastic pathway (Figure 4.2A).

Water can also cross the root epidermis and cortex by moving through the interior of the cells, which is known as the *symplastic* pathway (Figure 4.2). The internal contents of a cell, or *cytoplasm*, can be connected from one cell to the next by means of small tunnellike structures called *plasmodesmata* (singular, *plasmodesma*; Figure 4.2B). Water can move from cell to cell through the plasmodesmata without crossing the *cell membrane* surrounding the cytoplasm. The cell membrane is a formidable barrier to the movement of molecules, especially those that are electrically charged or are relatively large. Plasmodesmata therefore

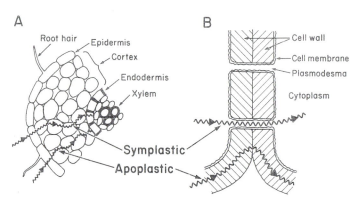

Figure 4.2   Anatomy associated with water uptake: (A) cross section of approximately one-quarter of a young root; and (B) detail of contact region between two adjacent cortical cells in a root. The apoplastic pathway in the cell walls and the symplastic pathway through the cell interiors are shown, but details of the structure within the two plasmodesmata are not.

greatly facilitate the diffusion of substances between adjacent cells, including water, which can move from cell to cell across the root cortex in the symplastic pathway (Figure 4.2).

At the layer of cells known as the *endodermis* (Figure 4.2A), the apoplastic pathway is blocked. Specifically, the cell walls where adjacent endodermal cells contact each other are filled with a waxy substance, *suberin*, which replaces the water between the cellulose polymers. Thus water and solutes from the soil must enter the cytoplasm of root endodermal cells in order to continue their movement to the vascular system of the root and eventually to reach the shoot. The endodermis is hence a site where root cells can regulate what enters the shoot.

After crossing the endodermis, water can enter the conducting cells of the vascular tissue known as the *xylem* (Figure 4.2A). Cells of mature xylem are simply hollow pipes made of cell walls. These "cells" lack a membrane, cytoplasm, and nuclei, and the cell walls between sequential cells along the xylem often disintegrate. This curious situation means that the water-conducting cells of the xylem function only when they are dead! Because these cells line up in a pipelike configuration, water and dissolved nutrients generally flow readily in the xylem along the root axis to reach the shoot.

Water Potential: Hydraulic Conductivity

When water flows downhill or over a waterfall, it is moving toward regions of lower energy, and so to analyze the movement of water in the soil and eventually into a root, we need to describe the energy of the water. In plant physiology, this energy is conventionally expressed per volume of water. Energy per volume has the same units as force per area, or pressure. We usually express the energy of water in terms of a quantity known as the *water potential*, which is represented by the capital Greek letter psi (Ψ) and has the units of pressure. The unit of pressure commonly used in plant physiology is the megapascal. A pressure of 1 atmosphere, or 760 millimeters of mercury, equals 0.101 megapascal, so 1 megapascal equals 9.9 atmospheres. To provide a reference point, pure water at atmospheric pressure is defined as having a water potential of 0.0 mcgapascal.

The water potential Ψ represents the potential for water to move, such as to flow downhill. The presence of dissolved solutes and of surfaces to which water can adhere lowers the energy of water. Thus the water potential in soil (Ψ_{soil}) and in plants, both of which contain solutes and have many internal surfaces, is less than 0.0 megapascal. Water spontaneously flows toward regions with lower values of Ψ. For water to move from the soil to the roots of plants, the water potential of the root, Ψ_{root}, must therefore be lower, or more negative, than Ψ_{soil}.

Rainfall raises Ψ_{soil} toward zero, beginning at the soil surface. The wetting front in the soil then moves downward, bringing moisture to the soil adjacent to roots. A rainfall of 10 millimeters (0.4 inch) on a dry sandy soil initially wets the uppermost part of the soil where the roots of agaves and cacti seldom occur. In about 10 hours, the wetting front for this rainfall moves downward to 9 centimeters (3.5 inches) in the soil, the mean depth of the roots for *Agave deserti* and *Ferocactus acanthodes* (Figure 4.1). There, Ψ_{soil} can be raised to approximately -0.4 megapascal, which is usually higher than Ψ_{root} for agaves and cacti that have been subjected to drought for a few weeks. Water then moves into the roots by flowing toward a region of lower water potential.

To quantify the movement of water, we use the hydraulic conductivities of the soil and the root, which indicate how easily water moves through these parts. We define *hydraulic conductivity* as the rate of water flow divided by the drop in water potential causing the flow. The greater the hydraulic conductivity, the greater is the tendency for water to flow. For an extremely dry sandy soil with a Ψ_{soil} of -10 megapascals, the hydraulic conductivity is extremely low. The water content of such a soil is also extremely low, only about 1 percent by volume (Figure 4.3). As the water content of a sandy soil increases to about 5 percent, Ψ_{soil} increases to about -1.0 megapascal, and the soil hydraulic conductivity increases about 1000-fold. If the soil is now made wet with about 25 percent water by volume, the soil water potential will become -0.1 megapascal or even higher, and the soil hydraulic conductivity will increase by another 1000-fold (Figure 4.3). Such wet soil is a good conductor of water, so water moves readily through it. Because of the high hydraulic conductivity of wet soil, water uptake into a plant in wet soil is restricted mainly by the hydraulic properties of the root.

The root hydraulic conductivity, generally given the symbol L_P and expressed in units of meter second^{-1} megapascal^{-1}, describes the ease of water entry into and flow along roots. The xylem cells of young roots of agaves and cacti less than about 7 days old have not matured, and so they have not yet become hollow pipes. Water cannot readily be conducted along such roots, and so L_P is relatively low. But as the xylem matures and the cells become hollow pipes made of cell walls, water is readily conducted axially along the roots, and L_P increases. When the roots are a month or so old, the waxy substance suberin found in the endodermis begins to accumulate in various other cell layers. Water entry from the soil into the roots then becomes more difficult, so L_P decreases during this phase of root aging for agaves. For cacti, this tendency for L_P to decrease with age because of suberin accumulation is compensated by an increase in the number of xylem conduits as the root thickens. This increase in conduit number facilitates the flow of water along the root axis. L_P has a maximal value of about 3×10^{-7} meter second^{-1} megapascal^{-1} at a root age of about 1 month for agaves and 12 months for cacti. The maximal L_P for roots of agaves and cacti is

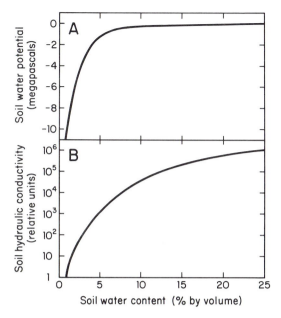

Figure 4.3 Influence of soil water content on (A) the soil water potential (Ψ_{soil}) and (B) the soil hydraulic conductivity. Data are for a sandy soil from the northwestern Sonoran Desert (Figure 1.5). For details, see Young and Nobel (1986).

similar to the maximal hydraulic conductivities for the roots of other taxa.

The soil water potential in the rooting region of agaves and cacti can be influenced by rocks in the soil as well as by large boulders at the soil surface. Rocks can divert water to their periphery and thereby greatly increase the soil water potential in certain regions of the soil. Flat horizontal rocks beneath the soil surface block the downward movement of water, leading to a higher Ψ_{soil} above such rocks after rainfall. During drought, the upper layers of the soil dry, and the lower wetter regions in the soil may then become the main source of water for agaves and cacti. As such water moves upward from the deeper wet regions toward the drier soil near the soil surface, its flow may be impeded by rocks, which can lead to relatively wet soil under rocks during drought.

The increase in Ψ_{soil} near rocks increases root branching and proliferation. The number of lateral roots per length of main roots for *A. deserti* is 11 times higher under rocks and 6 times higher alongside rocks than in rock-free regions. The roots of *Echinocereus engelmannii* are three times more profuse (indicated by root length per soil volume) near boulders than they are 5 centimeters (2 inches) away; those of *F. acanthodes* are five times more profuse under rocks than they are in rock-free regions; and the roots of *Opuntia acanthocarpa* are 12 times more profuse in the vicinity of rocks than they are in the dryer regions at the same depth in rock-free regions. Clearly, Ψ_{soil} is not constant at a particular depth in the soil. Such heterogeneities in water availability lead to regions in the soil that are crucial for seed germination as well as for water uptake by the roots of agaves and cacti.

Water Movement

The water potential of roots of well-watered agaves and cacti is about -0.3 megapascal. The soil must be wetter than this for water to be taken up. For a moderately dehydrated agave or cactus, as might occur after a few months of drought, the shoot and root water potential may decrease to -1.0 megapascal. Ψ_{soil} then need only be greater than -1.0 megapascal for water to be taken up.

When the soil is wet, water fills much of the pores between the soil particles and is in contact with the root surface. Water crosses the root epidermis and cortex by both the apoplastic and the symplastic pathways (Figure 4.2), with the symplastic pathway being more important for young roots of agaves and cacti. Water then crosses the endodermis and moves through mature xylem cells along the root to the shoot. At the onset of drought, however, the direction of the root–soil water flow is reversed. Water flow then is energetically favored from the roots to the surrounding drier soil. During drought, tiny air bubbles often form in the xylem, blocking water flow in the individual pipes just as air bubbles can block the flow of blood in capillaries. Indeed, both processes are called *embolism*. Suberization in various cell layers also generally increases during drought, which retards water movement between the root xylem and the root surface. Suberization and embolism lower L_P for roots of *Agave deserti*, *Ferocactus acanthodes*, and *Opuntia ficus-indica* by a factor of two to four during the first few weeks of drought.

Drought begins when Ψ_{soil} decreases below the root water potential, Ψ_{root}. The roots cannot then take up water, as water loss from the roots to the soil is energetically favored. The loss of water from root cells can cause a root to decrease in diameter and a root–soil air gap to develop. Two-month-old roots of *A. deserti*, *F. acanthodes*, and *O. ficus-indica* are about 2 millimeters (0.08 inch) in diameter. The diameter of these roots becomes an average of 10 percent less as Ψ_{soil} decreases from -0.1 megapascal to -1.0 megapascal and an additional 12 percent less as Ψ_{soil} decreases further to -10 megapascals. Compared with these 2-month-old roots, the amount of radial shrinkage is twice as much at 2 weeks of age and one-quarter as much at 12 months. For the young roots of various agronomic crops, the diameter can also shrink 20 to 60 percent in response to the soil's drying, similar to the responses of young roots of agaves and cacti. When roots shrink radially, the contact between the roots and the soil may be interrupted, and water must then diffuse in a vapor form to cross the resulting air gap. This slows the rate of water loss from the roots to a drying soil, which helps the plant conserve water. In addition, abscission of lateral roots during drought decreases the root area and hence can lessen the water loss for the entire root system, especially for agaves.

Water flows across the soil, the root–soil air gap, and the root sequentially. The hydraulic conductivity of the overall pathway from soil to root is influenced most by the part of the pathway that has the lowest, or most limiting, conductivity. For a plant in wet soil, the main limiter of water uptake is the root itself (Table 4.1). The drop in water potential in the soil toward the root surface is then relatively small, because the hydraulic conductivity of the wet soil is relatively high. Moreover, no root–soil air gap exists when the roots are fully hydrated. When the soil begins to dry, the hydraulic conductivity of the soil decreases (Figure 4.3), and the hydraulic properties of the root change very little. Nevertheless, root properties quantified by L_p are still the main limiter to water uptake during the initial phase of soil drying.

When drought commences, the roots begin to shrink radially, leading to a root–soil air gap. When such root shrinkage is about 10 percent or greater, the root–soil air gap can have the lowest hydraulic conductivity in the root–soil system (Table 4.1). This radial shrinkage of the roots during drought helps diminish the loss of water from the roots. As the soil dries further, Ψ_{soil} becomes less than -1.0 megapascal, as can occur 2 weeks after the beginning of drought for a sandy soil. The main factor limiting water movement then becomes the soil (Table 4.1). The soil hydraulic conductivity, which depends on the soil water content, decreases substantially as the soil dries. Over a period of about 1 month after the last rainfall, the soil conductivity can decrease by a factor of 1

Table 4.1   Effect of Soil Water Availability, as Quantified by the Water Potential Ψ, on the Hydraulic Conductivity That Most Influences Water Uptake or Loss by Roots of Agaves and Cacti Under Three Soil Conditions

	Soil condition		
Quantity	*Wet*	*Intermediate*	*Dry*
Soil water potential (megapascal)	Above -0.3	-0.3 to -1.0	Below -1.0
Water flow for root	Uptake	Variable	Loss
Limiting hydraulic conductivity	Root	Root–soil air gap (if root shrinks appreciably)	Soil

Note: For details, see Nobel and Cui (1992b).

million from wet sandy soil (Ψ_{soil} of -0.1 megapascal) to very dry sandy soil (Ψ_{soil} of -10 megapascals; Figure 4.3). The roots of agaves and cacti tend to occur in such porous soils, which quickly dry but also quickly rewet. Porous soils also allow the easy diffusion of oxygen toward the roots and carbon dioxide away from the roots.

Based on the various changes in hydraulic conductivities, the root–soil system acts as a rectifier for water movement into and out of roots. For wet soils, the overall conductivity is relatively high, and water uptake is limited only by the root hydraulic conductivity, L_P. Although L_P can decrease by a factor of two or more during drought, root shrinkage and especially the decrease in soil hydraulic conductivity play larger roles for decreasing the water flow out of roots. The rectification behavior of the soil can cause the overall conductivity of the root–soil system to decrease by a factor of more than 1000 during the first month of drought for sandy soils, and even more during prolonged drought. Decreases in soil hydraulic conductivity (Figure 4.3) are much larger than the decreases in root L_P during drought, and so changes in the soil can control water movement out of any root, even if L_P does not change. Consequently, the main mechanism for preventing an appreciable water loss from the roots of agaves and cacti to a drying soil lies in the properties of the soil.

What happens to the various hydraulic conductivities when the soil is rewet by rainfall after a drought? The changes in soil hydraulic conductivity can be completely and quickly reversed, and the soil no longer limits water movement (Table 4.1). As a root becomes rehydrated and swells, the root–soil air gap becomes progressively smaller. Full rehydration of root cells requires a few days following rewetting of the soil. During this period, the air gap surrounding the root may have the lowest, or most limiting, hydraulic conductivity in the root–soil system. Because the air bubbles that formed in the xylem during drought can be redissolved under wet conditions, water flow axially along a root can be fully restored. However, the accumulation of suberin in root cells during drought cannot be reversed, and so L_P for a root taken through a cycle of drought is lower than is L_P for a root under continuously wet conditions. On the other hand, rewetting can lead within 24 hours to the elongation of existing roots of agaves and cacti and the induction of new roots, especially new lateral roots. Because of the increase in root surface area by new root growth, the ability of the entire root system to take up water generally is greater a few weeks after rewetting than it was at the onset of drought.

Other Root Properties

In addition to taking up water, roots take up nutrients, help anchor plants in the soil, and can act as storage reservoirs for carbohydrates. Roots also interact with other organisms in the soil, including fungi, which in turn affect nutrient uptake by the roots. Nutrient levels in the soil influence nutrient levels in the shoots, which affect the nutritional value of agaves and cacti for animals. As we shall also see, the roots of agaves and cacti generally have a low tolerance of salinity, which has both agronomic and ecological implications.

Root:Shoot Ratios

The ability of CAM plants like agaves and cacti to conserve water is a characteristic that we will repeatedly encounter. Their stomatal opening at night when temperatures are lower leads to lower rates of water loss than for plants with daytime stomatal opening. Because of their lower water requirements, we might expect that agaves and cacti would have fewer roots than other plants do. On the other hand, a large root system would favor the collection of water, which is in scarce supply in a desert. A widely used measure for evaluating root behavior is the root:shoot ratio, which is the dry weight of the roots divided by the dry weight of the shoot.

Root:shoot ratios can be above 5 for shrubs from cold deserts (Table 4.2), although there are many exceptions. The high values suggest a large root system with a major storage role for these shrubs, many of whose shoots can experience considerable dieback during especially cold winters and hence need a carbohydrate supply for regrowth in the spring or summer. Root:shoot ratios average 0.9 for shrubs from warm deserts (Table 4.2). The root:shoot ratios of mature trees tend to be lower because of the large amount of woody tissue in the trunks; a typical

Table 4.2 Representative Root:Shoot Ratios
for Various Mature Perennials

Type	Root:shoot ratio
Cold desert shrubs	
Great Basin Desert	3.2 to 6.7
Commonwealth of Independent States (former Soviet Union)	3.6 to 7.3
Warm desert shrubs	
Chihuahuan Desert	0.6 to 1.3
Mojave Desert	0.5 to 1.1
Tropical and temperate trees	~0.3
Agaves	0.03 to 0.15
Cacti	0.08 to 0.14

Note: Data are based on dry weight. For details, see Nobel (1988) and Rundel and Nobel (1991).

value is about 0.3. The ratios for desert succulents are even lower, generally only about 0.1 (Table 4.2). For mature plants of *Agave deserti, Ferocactus acanthodes,* and *Opuntia ficus-indica,* the root:shoot ratio averages 0.10. Apparently, the lowest value reported for any plant species is 0.03 for *Agave lechuguilla* in the Chihuahuan Desert. Root:shoot ratios tend to be higher for younger agaves and cacti but in all cases are relatively low, except for species with massive taproots or large tuberous roots that store carbohydrates.

Because agaves and cacti are excellent water conservers and their shoots can store water, they apparently do not require a large root system. Their low root:shoot ratios also help minimize the loss of water to a drying soil. The length of roots per volume of soil tends to be less for agaves and cacti than for many other native species and for most cultivated crops. For this reason, less water per volume of soil tends to be lost from the roots to the soil during drought for agaves and cacti. Low root:shoot ratios also have implications for productivity (Chapter 7). In particular, less of the products made during photosynthesis has to be diverted from the shoots to the roots of agaves and cacti than is the case for plants with high root:shoot ratios (Table 4.2). On the other hand, their relatively small root system makes it rather easy to uproot agaves and cacti, a fact that is clearly recognized by illegal and legal collectors who readily remove such plants from the field. Also, barrel cacti are often knocked over by desert animals attempting to get to the water in their stems.

Mycorrhizae

The threads of fungi, referred to as *hyphae,* can form a closely woven mat around roots. They can even enter the roots between epidermal cells and can branch in the root cortex (Figure 4.2A). Such associations between fungal hyphae and roots are referred to as *mycorrhizae.* The associations are mutually beneficial, as the fungi supply nutrients to the plant and in return gain organic molecules, such as sugars. Mycorrhizae occur for various agaves and cacti, but they are not as prevalent for desert succulents as they are for many other taxa. Nevertheless, the uptake of phosphorus, a nutrient that is often at low levels in soils but is required in large amounts by plants, can be enhanced by the mycorrhizae of agaves and cacti. The fungal hyphae outside a root increase the area of contact with the soil compared with a root surface without hyphae. Root hairs (Figure 4.2A) similarly increase the surface area of roots in contact with the soil particles and can also increase nutrient uptake.

Specific mycorrhizal interactions have been examined for three species of agaves and cacti—*Agave deserti, Ferocactus acanthodes,* and *Opuntia ficus-indica.* In the northwestern Sonoran Desert (Figure 1.5), mycorrhizae occur along 2 to 11 percent of the root length of these succulents. This is a low level of infection compared with that of other desert plants and cultivated species, whose roots can be infected over 60 percent of their length. The infection level of the three species—especially for the lateral roots of A. deserti—can be raised by maintaining the plants under wet conditions.

Under wet conditions, mycorrhizae increase the shoot levels of phosphorus and zinc, the latter nutrient required in small amounts by plants. The mycorrhizal association for A. deserti can also increase L_P for water uptake by facilitating the movement of water from the root surface to the endodermis (Figure 4.2A). Compared with plants without such fungal associations, mycorrhizae for A. deserti under continuously wet conditions enhance water and nutrient uptake, leading to a 19 percent increase in carbon dioxide uptake by the shoot. The effects are probably smaller under field conditions, and indeed much remains to be learned about the importance of mycorrhizae for agaves and cacti.

Nutrients

The nutrient requirements and the nutrient contents of agaves and cacti are basically similar to those of other plants (Table 4.3). Nitrogen is often the soil element most limiting for growth, both agronomically and in natural environments. Fertilization with nitrate-containing inorganic fertilizers or manure almost always enhances the growth of cultivated agaves and cacti. *Opuntia ficus-indica* raised for nopalitos in Milpa Alta

Table 4.3 Average Element Levels in the Chlorenchyma of Various Taxa

Element level	Agaves	Cacti	Leafy agronomic plants
Macronutrients			
Nitrogen (%)	1.2	1.5	3
Phosphorus (ppm)	2,100	1,700	3,000
Potassium (%)	1.8	1.6	2.0
Calcium (%)	3.7	4.4	2.0
Magnesium (%)	0.7	1.1	0.7
Micronutrients			
Manganese (ppm)	30	140	70
Copper (ppm)	4	7	8
Zinc (ppm)	26	28	40
Iron (ppm)	77	130	150
Boron (ppm)	24	39	30
Sodium (ppm)	43	150	1,000

Note: Data are expressed as percent or as parts per million (ppm) by dry weight. For details, see Nobel (1988).

near Mexico City (Figure 3.3) is fertilized annually with a 20-centimeter (8-inch) layer of cow manure. The other two common nutrients in fertilizers are phosphorus and potassium. Application of these two elements can increase the growth of agaves and cacti, but the results depend on the element levels already in the soil.

Nutrient responses are extremely complicated and varied. Elements interact with each other in the soil and during uptake into roots. For instance, potassium interacts with sodium, calcium with magnesium, and copper with zinc. In addition, the amounts of most elements vary considerably in soil from different locations, as is the case for boron, which plants require in small amounts. Sufficient boron for the growth of agaves and cacti is present in the form of borate in the soils of the Sonoran Desert in the southwestern United States. However, the extremely low levels of boron in the Chihuahuan Desert of north-central Mexico can limit the growth of cacti.

Plant nutrients are generally divided into two categories based on the relative amounts required. *Macronutrients* are required in greater amounts than are *micronutrients* (Table 4.3). Among the macronutrients, agaves and cacti contain lower levels of nitrogen and phosphorus and a higher level of calcium than do leafy agronomic plants. The levels of nitrogen and phosphorus in platyopuntias are generally lower than are those required by cattle. Thus these two elements usually must be supplemented when cladodes are used for fodder, as mentioned in Chapter 3. Calcium in plants is often combined with organic molecules, and its high level in cactus stems can even deter some animals from eating cacti. Micronutrients are usually present in slightly lower levels in agaves and cacti than in most agronomic plants, but considerable variation exists (Table 4.3). Micronutrients are combined in various ways with organic molecules in cells and are crucial in small amounts for the functioning of essentially all metabolic pathways.

Nutrients can simply diffuse into roots, but more often they are transported into roots by an energy-requiring process known as *active transport*. Very little is known about the active transport of macronutrients and micronutrients into the roots of agaves and cacti. Such nutrients move across the root cortex, presumably mainly in the symplastic pathway (Figure 4.2), before entering the endodermal cells and eventually the root xylem. The nutrients then move along the root xylem to reach the shoot.

Salinity

Most agaves and cacti do not tolerate even moderate levels of soil salinity. Growth of their roots is usually drastically inhibited by levels of sodium at one-fifth the concentration found in seawater (seawater contains about 12 grams [0.4 ounce] of sodium per liter [1 liter = 1.06

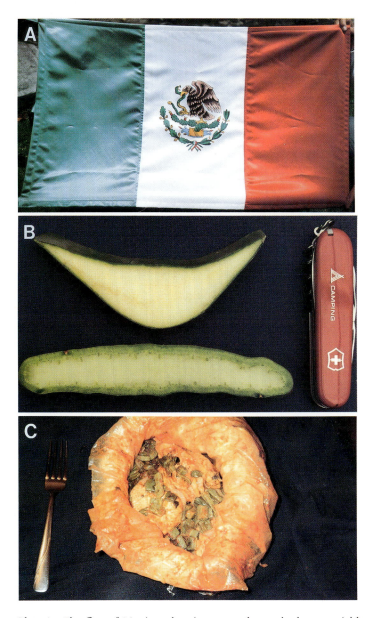

Plate A   The flag of Mexico, showing an eagle perched on a prickly pear cactus, possibly *Opuntia streptacantha*. Its design comes from an Aztec legend about a signal for reaching the promised land (present-day Mexico City) in 1325.

Plate B   Cross section of a leaf of *Agave americana* (upper) and a cladode of *Opuntia ficus-indica* (lower) showing the green chlorenchyma, where photosynthesis takes place, and the whitish water-storage parenchyma, which acts as a water reservoir. The chlorenchyma averages 1 millimeter in thickness for the agave and 4 millimeters for the cactus.

Plate C   Mixiote, a delicious dish of chicken, nopalitos (diced young cladodes of *Opuntia ficus-indica*), other vegetables, and spices that are wrapped in the cuticle of *Agave mapisaga* or *A. salmiana* for cooking. The cuticle has been pulled back to show the contents.

Plate D   Two workers emptying aguamiel that they have just sucked from basins created in *Agave mapisaga* (Figure 2.3) into the gourds. The wooden barrels, each of which can contain 30 liters, are transported to a tinacal for fermentation into pulque.

Plate E   Dispensing pulque to a child sent by her grandfather. The sites are near Calpulalpan, Tlaxcala, Mexico.

Plate F   A variegated form of a horticulturally important agave, *Agave americana*.

Plate G   *Agave filifera*.

Plate H   Cactus pears from a popular variety of *Opuntia ficus-indica* purchased in a Los Angeles, California, supermarket, one of which has been cut open to expose the red pulp.

Plate I   A female cochineal insect (*Dactylopius opuntiae*) being crushed to produce a reddish purple fluid. Note the insect body on the finger and the heavily infested cladode of *O. ficus-indica* in the background.

Plate J   Singeing spines off a prickly pear cactus (*Opuntia engelmannii*) on the Maltsberger Ranch, Cotulla, Texas. (Photograph is courtesy of William A. Maltsberger.)

Plate K   Flowers of *Coryphantha vivipara* variety *rosea* in the laboratory.

Plate L   Flower of an *Epiphyllum* hybrid, a group important to the horticultural trade. The site is Rainbow Gardens, Vista, California.

quarts]). Very little sodium moves into the shoots of cacti, so cladodes used as animal fodder must be supplemented with sodium chloride. Indeed, the levels of sodium in the shoots of agaves and cacti are much lower than those in leafy agronomic plants (Table 4.3; note that sodium is not considered a nutrient for most plants).

The high sensitivity of agaves and cacti to salinity has many consequences. When cacti grown for fruit are watered by drip irrigation, care must be taken to avoid the accumulation of sodium in the rooting zone. One solution is to use a rootstock that can tolerate salinity, such as *Opuntia quimilo* from Argentina. Another cactus, *Cereus validus*, is native to regions in Argentina that have salt flats in the dry season. Evidently, its roots succumb to the salinity and abscise during drought. Rain during the next wet season dilutes the sodium and induces new main roots, which in turn are shed during the next dry season.

Cereus validus is more tolerant of salinity than are most agaves and cacti. Exposure to half the concentration of sodium chloride in seawater for 2 weeks has minimal effects on its growth, whereas one-quarter this concentration for 2 weeks approximately halves the growth of seedlings of *Agave deserti*. Continuous exposure to one-tenth the concentration of sodium chloride in seawater halves the growth of *Opuntia ficus-indica*. The low tolerances to salinity somewhat restrict the regions where agaves and cacti can be successfully cultivated, although most soils are not saline. This restriction can be removed by selecting tolerant types and by breeding efforts, which are clearly warranted by the high potential productivity and commercial importance of agaves and cacti.

REFERENCES

Cannon, W. A. 1911. *The Root Habits of Desert Plants.* Carnegie Institution of Washington, Washington, D.C.

Cui, M., and P. S. Nobel. 1992. Nutrient status, water uptake, and gas exchange for three desert succulents infected with mycorrhizae fungi. *The New Phytologist* 122:643–649.

Hunt, E. R., Jr., and P. S. Nobel. 1987. A two-dimensional model for water uptake by desert succulents: Implications of root distribution. *Annals of Botany* 59:559–569.

Jordan, P. W., and P. S. Nobel. 1984. Thermal and water relations of roots of desert succulents. *Annals of Botany* 54:705–717.

Nobel, P. S. 1988. *Environmental Biology of Agaves and Cacti.* Cambridge University Press, New York.

Nobel, P. S., and M. Cui. 1992a. Prediction and measurement of gap water vapor conductance for roots located concentrically and eccentrically in air gaps. *Plant and Soil* 145:157–166.

Nobel, P. S., and M. Cui. 1992b. Shrinkage of attached roots of *Opuntia ficus-*

indica in response to lowered water potentials—predicted consequences for water uptake or loss to soil. *Annals of Botany* 70:485–491.

Nobel, P. S., P. M. Miller, and E. A. Graham. 1992. Influence of rocks on soil temperature, soil water potential, and rooting patterns for desert succulents. *Oecologia* 92:90–96.

Nobel, P. S., and J. Sanderson. 1984. Rectifier-like activities of roots of two desert succulents. *Journal of Experimental Botany* 35:727–737.

Rundel, P. W., and P. S. Nobel. 1991. Structure and function in desert root systems. In *Plant Root Growth: An Ecological Perspective* (D. Atkinson, ed.). Blackwell Scientific, Oxford. Pp. 349–378.

Young, D. R., and P. S. Nobel. 1986. Predictions of soil–water potentials in the north-western Sonoran Desert. *Journal of Ecology* 74:143–154.

5

Shoots: Environmental Interactions

Water conservation is at the very heart of the ecological advantages and the agronomic potential of agaves and cacti. Their shoots typically store large volumes of water in relation to their surface area across which water is lost by transpiration. These large volume:area ratios of agaves and cacti are a crucial adaptation to arid and semiarid regions. The shoots of agaves and cacti are also interesting in their adaptations to temperature and solar radiation. The top, or apex, of a barrel cactus like *Ferocactus acanthodes* has a dense covering of spines and hairs (Figure 1.7A) that influence the temperature of the dividing cells in the apical meristem. In addition, spines affect the amount of light that reaches the stem surface and that hence is available for photosynthesis. The size and orientation of agave leaves and cactus stems also influences radiation interception, photosynthesis, growth, and shoot temperature.

Our main objective in this chapter is to interpret the morphologies of agaves and cacti with respect to water use, radiation interception, and responses to temperature. Some of these properties are related to survival, including adaptations to avoid freezing damage. To understand freezing damage, we will examine cellular properties, because the death of tissues and shoots begins with the death of cells. Interception of solar radiation is necessary for photosynthesis but can raise shoot temperatures, so we will also consider the high-temperature tolerances of agaves and cacti.

Water Relations

The expression "water relations" refers to a wide range of processes involving water and plants, beginning with water uptake from the soil and ending with its loss from the shoot. Water and nutrients from the soil move through the root xylem and into the shoot xylem to reach the shoot cells. Water is necessary for cell expansion, which relates to both growth and water storage. Biochemical processes take place in water, as in the cytoplasm of plant cells. The growth of plants requires stomatal opening for carbon dioxide uptake. This is accompanied by the inevitable loss of water from the plants as it evaporates and diffuses from its high concentration in the shoots to a lower concentration in the surrounding air. The evaporation of water inside shoots is a cooling process, so transpiration lowers the plant's temperature. Not surprisingly, water is involved in essentially every aspect of plant physiology.

The shoots of agaves and cacti generally have a high water content. Indeed, barrel cacti have long been a source of water for native animals as well as thirsty desert travelers. We can quantify the high shoot water content by using the water potential, Ψ, which we introduced in the previous chapter. Shoots of agaves and cacti generally have a Ψ of -0.3 to -0.5 megapascal when hydrated under conditions of wet soil. Even after 4 to 6 months of drought, Ψ does not drop much below -1.0 megapascal for most agaves and cacti. On the other hand, Ψ of leaves of most agronomic plants often decreases below -1.0 megapascal on a daily basis, even when the soil is wet. Agaves and cacti have relatively small daily and seasonal fluctuations in Ψ because of their low water-loss rates due to nocturnal stomatal opening, their large capacity for water storage, and the low water permeability of the epidermis with its waxy cuticle.

Water Storage

The leaves of agaves and stems of most cacti generally have large amounts of water-storage parenchyma (Plate B). This whitish tissue can store appreciable amounts of water and thereby helps the plants survive long periods of drought. During drought, water can move from the water-storage parenchyma to the chlorenchyma, where photosynthesis and other crucial metabolic activities take place. The chlorenchyma is also relatively thick and provides additional water storage for certain agaves and cacti (Plate B).

The water-storage capability of shoots can be quantified by their volume:area ratio. This ratio, which has the dimension of length, represents the average depth of the tissue from which water can be obtained. For the relatively thin flat leaves of most plants, the volume:area ratio is one-half of the leaf thickness, averaging only 0.016 centimeter (0.006 inch; Table 5.1). When removed from the plant, such leaves rapidly lose water and wilt. The leaves of agaves and the cladodes of opuntias generally have mean volume:area ratios of about 1 centimeter, indicating a much greater depth of stored water than for thin leaves. The massive stems of columnar cacti and barrel cacti can have volume:area ratios that exceed 5 centimeters (2 inches; Table 5.1). These species contain an enormous volume of stored water in their stems, compared with the

Table 5.1 Volume:Area Ratios for Various Leaves and Stems

Organ	Volume:area (centimeters)*	Seedling	Volume:area (centimeters)*
Leaves of garden vege-tables	0.010 to 0.022	*Agave deserti* 2 weeks old	0.07
Leaves of deciduous trees	0.013 to 0.020	20 weeks old *Ferocactus acanthodes*	0.18
Leaves of agaves	0.3 to 2.0	1 week old	0.02
Cladodes of opuntias	0.4 to 1.5	8 weeks old	0.06
Stems of columnar cacti	2 to 6	1 year old	0.17
Stems of barrel cacti	4 to 9		

Note: Data are for mature plants, except where indicated. For details and references, see Nobel (1988).

*Ratios were calculated from the volume of the organ or the seedling shoot divided by its total surface area (area of both sides for leaves). One centimeter equals 0.4 inch.

surface area across which water can be lost to the atmosphere. Such plants can tolerate prolonged drought, ranging from months to even years.

As seedlings of agaves grow older, their leaves become thicker, and as seedlings of cacti grow older, the juvenile leaves are shed and the stems become thicker. The volume:area ratios of seedlings of both agaves and cacti thus increase with age (Table 5.1). A 2-week-old seedling of *Agave deserti* (Figure 1.5) can tolerate a drought of about 2 weeks. But a 20-week-old seedling can tolerate a drought seven times longer because its leaves are thicker and the depth within the leaf from which water can be obtained has increased (changes also occur in the epidermis over this period, as we will consider shortly). The volume:area ratio of a sphere equals one-third of its radius (Chapter 1) and so increases as the sphere gets larger. For the approximately spherical seedlings of *Ferocactus acanthodes*, the volume:area ratio increases nearly 10-fold from 1 week to 1 year of age (Table 5.1). When plants of this species are maintained in wet soil for 1 year, their stems can be 1 centimeter in diameter and can tolerate a drought of 4 to 6 months.

The Epidermis

The epidermis controls water loss from the shoot and can differ for agaves and cacti, compared with most other plants. It is covered by a relatively waterproof cuticle and is interrupted by stomatal pores through which water vapor and carbon dioxide can readily diffuse (Figure 1.8). Thicker cuticles and fewer stomates per unit area make the movement of gases across the epidermis more difficult, which is critical for the water relations of agaves and cacti.

The scanning electron micrographs in Figure 1.8 indicate that there are about 35 stomates per square millimeter of epidermis for leaves of *Agave attenuata* and stems of *Opuntia ficus-indica* (Figure 1.7C). This number of stomates per unit area is typical for agaves and cacti but is much lower than the average values for thin leaves of other species (except for the upper leaf surfaces of most dicotyledons, which have few or no stomates; Table 5.2). The greater number of stomates per shoot area for thin leaves correlates with their higher rates of transpiration, compared with those of agaves and cacti under similar conditions.

The cuticle tends to be much thicker for agaves and cacti than for species with thin leaves (Table 5.2). Cuticles of agaves and cacti are often 20 micrometers or more in thickness (1 micrometer = 10^{-6} meter = 0.00004 inch). The cuticle tends to thicken as the plant grows older, although cracks can develop in the cuticle of older shoots. When the stomates are open, water vapor, carbon dioxide, and oxygen readily pass in and out through the stomatal pores, and relatively little movement of these gases occurs across the cuticle. As the stomates partially close, the exchange of gases with the surrounding air decreases. If the stomates close completely, gas exchange between the inside of leaves or stems and the environment is greatly reduced, because the only pathway remaining is across the cuticle. The thick cuticle of agaves and cacti thus helps to ensure that little water vapor diffuses out of their shoots during periods of stomatal closure, which is a major factor enabling them to tolerate long periods of drought in arid and semi-arid regions.

Table 5.2 Representative Epidermal Properties for Various Leaves and Stems

Organ	Number of stomates per square millimeter*	Cuticle thickness (micrometers)*
Leaves of monocotyledonous crops	50 to 200	0.2 to 2
Leaves of dicotyledonous crops	10 to 200	0.2 to 3
Leaves of deciduous trees	0 to 400	1 to 6
Leaves of agaves	20 to 50	3 to 20
Stems of cacti	15 to 60	3 to 40

Note: For details, see Martin and Juniper (1970), Nobel (1988), and Fahn (1990).

*The low end of the ranges for dicotyledons and deciduous trees refers to their upper leaf surfaces. One millimeter equals 1000 micrometers equals 0.04 inch.

Drought Tolerance

The ability of agaves and cacti to tolerate drought is the result of a combination of features, including shoot water storage, the low water permeability of the epidermis, and the ability of cells to withstand dehydration. We have already indicated that larger seedlings can tolerate longer droughts, in part because of higher volume:area ratios. The epidermis of agaves and cacti tends to become less permeable to water as the cuticle thickens with age. Mature plants can tolerate droughts of many months to even years. For example, *Copiapoa cinerea* in the Atacama Desert of coastal northern Chile (Figure 1.6) grows in regions where the annual rainfall is only 10 millimeters (0.4 inch). This relatively small barrel cactus has survived rainless periods of 6 years!

During prolonged drought, considerable water is lost from the water-storage parenchyma (Plate B) and from other shoot cells. The loss of water causes the cells to shrink and their cell walls to become more irregular in appearance (Figure 5.1). Air gaps generally develop between the cells during drought. As water continues to be lost, air-filled gaps can also develop within the cells when the cell membrane pulls away from the cell wall. The concentration of dissolved solutes inside the cells increases as water is lost. That is, the same number of molecules per cell are dissolved in a much smaller volume of water as drought progresses (Figure 5.1). In turn, high concentrations of solutes can adversely affect various cellular processes by disrupting molecular structure, membranes, and chemical reactions.

For the thin leaves of most species, a water loss of 30 percent is

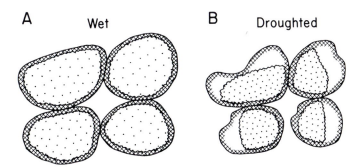

Figure 5.1   Cells in the water-storage parenchyma (A) under well-watered conditions and (B) after prolonged drought. During drought, the loss of intracellular water causes the cell membrane (wavy line) surrounding the cytoplasm to pull away from the cell wall (hatched area). This leads to a higher concentration of cellular solutes (represented by a greater density of the stippling).

usually lethal. In contrast, cells of the water-storage parenchyma of many agaves and cacti can tolerate water losses of 70 to 95 percent by volume before sustaining irreversible damage. The cells of these succulent plants evidently can withstand a substantial concentrating of solutes and appreciable changes in cell shape (Figure 5.1) before their cellular processes are disrupted. During the early phases of drought, the water-storage parenchyma can decrease 30 to 40 percent in thickness, while the chlorenchyma decreases less than 10 percent (Plate B). Water is hence lost preferentially from the water-storage parenchyma rather than from the chlorenchyma. Both the preferential loss of water from water-storage tissues and the tolerance of extreme cellular dehydration are important adaptations of agaves and cacti to arid environments.

Radiation

The term *radiation* is often used synonymously with the word *light*. Light is composed of photons, some of which can be absorbed by photosynthetic pigments in chloroplasts, especially the green pigment chlorophyll. For such considerations, the amount of such light striking a surface area per time is often called the *photosynthetic photon flux* (PPF). PPF refers to only those photons with wavelengths (colors) in the visible region of the spectrum, the region where chlorophyll absorbs the photons used for photosynthesis. Chlorophyll strongly absorbs blue light and red light and mostly reflects wavelengths between red and blue, such as green. Therefore, leaves and stems of plants generally appear greenish. When relating radiation to the temperatures of plants, we must refer to the energy carried by photons of all wavelengths.

Light Distribution over Agaves

The shape of the agave shoot is ideal for distributing light relatively uniformly over all its leaves. Generally, the more uniform the distribution of light over the leaf surfaces, the greater is the photosynthesis for the plant as a whole. As we discussed in Chapter 1 in relation to Fibonacci numbers, a new leaf of an agave unfolds from the central cone of folded leaves at 137° clockwise or counterclockwise from the previously unfolded leaf. This unfolding pattern leads to a rosette of leaves pointing in different directions at regular intervals. The most recently unfolded leaves are nearly vertical, and the oldest leaves are nearly horizontal. For most of the day, sunlight consequently can be absorbed by the younger leaves and also can reach the older (lower) leaves.

To help appreciate the advantages of the rosette leaf arrangement for agaves with respect to the absorption of the photosynthetic photon flux (PPF), let us imagine an upside-down agave. For such an orientation, light would be intercepted predominantly by the older, horizontal

leaves, and so little light would be available for the nearly vertical, younger leaves, which for an upside-down agave would be shaded by the older leaves. For the right-side-up agave, the regular unfolding of leaves at 137° intervals minimizes interleaf shading and helps ensure a relatively uniform distribution of light over the leaves. The PPF is thus not too high on some leaves and too low on others. Rather, the photons arrive at such a rate that nearly all of them can be used effectively for photosynthesis. This maximizes photosynthesis and growth for the plant as a whole.

Cactus Morphology

Many morphological features of cacti influence the PPF that reaches the surface of their stems. Spines shade the stem, decreasing the light available for photosynthesis. The spines of the teddy bear cholla (*Opuntia bigelovii*) reduce the PPF incident on the stem by 30 to 40 percent. Periodically trimming its spines can lead to a 50 percent stimulation of growth (measured by increases in stem volume) over a 2.5-year period. Similarly, spines shade the stems of various barrel cacti and so diminish their growth. There is thus a trade-off between growth and the protection provided by spines against thermal extremes and herbivory.

Pubescence, or the "hairiness" caused by the hairlike cells projecting from the shoot epidermis, also reduces the light incident on stems. The apical regions of many barrel cacti and columnar cacti have much pubescence. The regions near the top of stems of cacti in the genera *Backebergia, Cephalocereus, Discocactus, Lophocereus,* and *Melocactus* as well as other cacti also have much pubescence and many spines. The resulting shading leads to a lower rate of photosynthesis. The branches, or arms, on the saguaro (*Carnegiea gigantea*; Figure 1.7B) can shade the main stem, and vice versa. The orientation of stems, such as the equatorial tilting of certain barrel cacti, also affects PPF interception and the resulting growth of cacti. Ribs and other three-dimensional features of the stem surface increase the stem area and influence the PPF distribution on the stem. Ribs are also involved in the water relations of barrel and columnar cacti, allowing the stems to swell, like the pleats of an accordion, with the uptake of soil water after rainfall.

Cladode Orientation

Numerous measurements of the orientation of cladodes of more than 20 species of platyopuntias in the late 1970s and early 1980s indicated nonrandom patterns. These patterns are usually not obvious visually. If a class of students were to form a ring around a group of platyopuntias and were asked to raise their hands if they thought the majority of the cladodes faced toward them, nearly all the students would raise their hands. This is an illusion created by the much greater area of the flat sides of a cladode compared with its narrow edge. A cladode seen face-on thus makes a much greater visual impression than does one seen edge-on. Therefore, to observe patterns, the orientations of many un-shaded, vertical cladodes on neighboring plants must be determined using an accurate compass. In many cases, the sides of at least twice as many unshaded cladodes face within 10 degrees of east–west compared with facing within 10 degrees of north–south (Figure 5.2). However, at latitudes more than about 30° from the equator, cladodes sometimes tend to face north–south (Figure 5.2).

The observed orientation of cladodes maximizes the interception of the photosynthetic photon flux (PPF) at times of the year when new cladodes are initiated. At nearly all latitudes where platyopuntias occur naturally or are cultivated, an unshaded vertical surface facing east–west intercepts more light throughout the year than does a surface at any other orientation. Consistent with this, unshaded cladodes tend to face east–west at nearly all sites. However, facing north–south can become advantageous in the winter at latitudes about 30° or more from the equator. The sun's trajectory is then low enough in the sky to allow appreciable PPF to strike the equatorial-facing side over the course of a

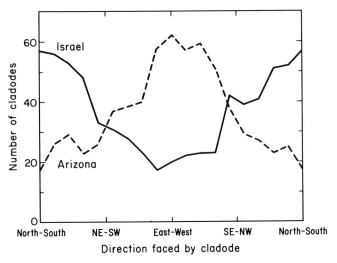

Figure 5.2   Orientation of cladodes of *Opuntia phaeacantha* in Arizona and in Israel. The sites are at 32° north latitude in the central Sonoran Desert in Arizona, a region of predominantly summer rainfall, and at 31° north latitude in Israel, a region of predominantly winter rainfall. At each site, the orientations of 660 unshaded, terminal cladodes were measured and assigned to angle classes 10 degrees wide centered on the direction faced (NE refers to northeast; SW, to southwest). For details, see Nobel (1982).

day. The polar-facing side gets relatively little PPF, but the sum of the PPF for the north plus the south sides is then more than for any other orientation of a vertical cladode. Cladodes of *Opuntia ficus-indica* and *O. phaeacantha* (Figure 5.2) tend to face north–south in regions of winter rainfall in northern Israel, as do those initiated in the winter in California and central Chile. Not incidentally, all these locations are more than 30° from the equator.

The mechanism underlying the orientation tendency of cladodes illustrates an interplay between morphology and physiology. The "daughter" cladodes of most platyopuntias tend to occur in the same plane as the "mother" cladodes on which they originate. If the mother cladode is oriented in a direction favorable for intercepting PPF, such a cladode can perform more photosynthesis and therefore tends to produce more daughter cladodes (and more flowers). Because the daughter cladodes tend to be aligned in the same plane as the underlying mother cladodes, they have the same favorable orientation. PPF interception by the whole plant and thus biomass productivity are thereby enhanced, compared with plants with a random orientation of daughter cladodes. The orientation tendency of the daughter cladodes can even be affected by shifts in the prevailing direction of light caused by mountains, other topographic features, shading by neighboring plants, and shading by other cladodes on the same plant.

Although mature cladodes have fixed orientations, a slight rotation is possible during development for very young cladodes. Daughter cladodes of *O. ficus-indica* can rotate up to 16° toward a more favorable orientation (with respect to the prevailing light) compared with the direction they faced when initiated. Such *phototropic* responses increase PPF interception by the daughter cladodes. Another phototropic response is the equatorial tilting of the stems of barrel cacti like *Ferocactus wislizenii* in North America and *Copiapoa cinerea* in South America (Figure 1.6). This orientation leads to a greater radiation interception at the top of the stem, which raises the temperature there and warms the apical meristem on sunny but cold days. Increases in temperature can also enhance flowering and can speed up reproduction.

Temperature Relations

Temperature affects nearly all processes in plants, so factors that affect tissue temperature are important for the physiology of agaves and cacti. The temperature of a particular part of a shoot depends on the morphology of that part and the local environmental conditions, such as the level of incident radiation. The sunlit side of a barrel cactus can be much warmer than the opposite side. Plant temperature is also affected by the rate of transpiration. Water loss involves the evaporation of water, which is a cooling process, so transpiration lowers tissue temperatures.

The main environmental influence on plant temperature, however, is the temperature of the surrounding air. Plant parts, especially thin leaves or small stems, tend to be closer to air temperature as the wind speed increases. Temperature extremes profoundly affect plant survival and distribution, so we will also consider the tolerances of agaves and cacti to both low temperatures and high temperatures.

Morphology

Plant shape and surface characteristics have major influences on the interception of radiation and hence on tissue temperature. We begin by considering the influence of the spines and pubescence, which occur near the top of the barrel cactus *Ferocactus acanthodes* (Figure 1.7A). Just as a hat can shield us from the sun or warm us during a cold night, coverings of the tops of cactus stems lower the maximal temperatures and raise the minimal temperatures of the apical meristem, where the precursor cells for the entire stem originate.

In the absence of spines and pubescence, the apex of the stem of the saguaro (*Carnegiea gigantea*) may be 50°C (122°F) on a warm summer day (Figure 5.3A). As the spine coverage increases, the shading of the apex lowers the maximal tissue temperature. Complete shading of the apex by spines reduces the predicted maximal apical temperature by 10°C (18°F). Similarly, as the thickness of the layer of apical pubescence increases, the maximal temperature also decreases. The 10 millimeters (0.4 inch) of pubescence that occurs for some barrel and columnar cacti reduces the predicted maximal temperature during the summertime by 12°C. In most cases, there is a combination of both spines and pubescence at the apex of barrel and columnar cacti, and together they can reduce the maximal apical temperature by 14°C (Figure 5.3A). Cellular processes and ultimately the survival of the apical meristem depend on such large reductions of maximal temperatures during episodes of high temperature.

Spines and pubescence can moderate minimal temperature as well. In the absence of apical coverage, tissues at the top of a columnar cactus might reach −10°C (14°F) at night during the winter (Figure 5.3B), which is considerably below freezing. Spines and pubescence then act like an insulating hat, raising the tissue temperatures above those that may damage the cells. Complete shading of the apex by spines or a 10-millimeter thickness of apical pubescence can raise the predicted temperature to 0°C (32°F; Figure 5.3B). This avoids the freezing temperature of −10°C that could occur in the absence of apical protection and hence reduces the possibility of low-temperature damage.

Apical spines and pubescence vary greatly for cacti in nature. Among the four species of barrel cacti in the southwestern United States, *Ferocactus viridescens* grows in the least cold habitats along coastal regions of

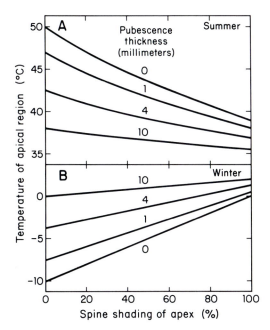

Figure 5.3 Influence of spines and pubescence on the temperature of the apical meristem at the top of saguaro, *Carnegiea gigantea*: (A) maximal temperatures on a summer day; and (B) minimal temperatures on a winter night. One millimeter equals 0.04 inch. Data are based on a computer simulation model, as described in Nobel (1988).

southern California. Consistent with its distribution, *F. viridescens* has only 2 millimeters of apical pubescence and a 22 percent shading of the apex by spines. At the other extreme, *Ferocactus acanthodes* growing at cold sites at an elevation of 1600 meters (5200 feet) in southern Nevada has 9 millimeters of apical pubescence and 93 percent shading of the apex by spines. *Ferocactus covillei* and *F. wislizenii* are intermediate in apical pubescence and spine shading and can be found in regions of Arizona that are intermediate in minimal wintertime temperatures compared with those of the other two species.

The apical pubescence for saguaro (*C. gigantea*) is usually quite thick, 9 to 10 millimeters, but its spine shading of the apex varies with the minimal wintertime temperatures in different parts of its distribution. In the southern part of its distribution in Sonora, Mexico, spine shading for saguaro is less than 10 percent, increasing to over 40 percent in the northern part of its distribution at similar elevations in central Arizona. For a common barrel cactus in Chile, *Eriosyce ceratistes*, apical pubescence remains at 7 millimeters in thickness, but spine shading of the apex increases from 23 to 67 percent along a distance of about 400 kilometers (250 miles) away from the equator (southward). Based on a computer model that estimates the influence of morphology on apical temperature (Figure 5.3), the upper elevational limits of *E. ceratistes* in the Andes can be predicted to within 30 meters (100 feet) of its measured elevational limit. The upper limits of the columnar cactus *Trichocereus chilensis* (Figure 3.7), which also occurs at over 2000 meters (6600 feet) in the Andes of Chile, can be equally well predicted over a somewhat greater range in latitude. The upper elevations for these two species are thus apparently determined by low temperatures at the plant apex during the winter.

Although the influences of spines and pubescence on stem temperature are particularly well understood, other morphological features also affect tissue temperature. Thin agave leaves and cactus stems tend to be closer to air temperature than thicker ones. The temperature of massive stems of barrel cacti can be very different from the air temperature. In the morning, their sunlit eastern sides can be considerably above the temperature of the surrounding air, sometimes by 20°C (36°F), while their western sides are below the air temperature. For these massive cacti, transpirational cooling at night can lower the tissue temperatures near the stem surface by only 1°C to 2°C. For the much thinner leaves of most agaves and the thinner cladodes of opuntias, transpiration at night can lower chlorenchyma temperatures by 3°C to 4°C (5°F to 7°F). During the daytime when the stomates of CAM plants tend to be closed, transpiration has little effect on shoot temperatures of agaves and cacti.

The color of the plants also affects the absorption of sunlight and thus their daytime temperature. Dark green leaves or stems with large amounts of chlorophyll per area tend to be about 3°C warmer in direct sunlight than pale green shoots of the same shape. In addition, cuticles often reflect considerable sunlight, causing leaves and stems to appear light grayish green and leading to lower shoot temperatures. Many factors thus influence the chlorenchyma temperature of agaves and cacti.

Low Temperature

Some agaves and many cacti can tolerate air temperatures substantially below freezing. The three agaves with the greatest tolerance to low temperature are *Agave deserti* (Figure 1.5), *A. parryi*, and *A. utahensis*, all of which can tolerate −20°C (−4°F). *Opuntia fragilis* and *O. humifusa* are platyopuntias native to Canada, with *O. fragilis* occurring up to 58° north latitude in northern Alberta, where temperatures can reach −40°C (−40°F)! These and other cactus species from high elevations in the United States, such as *Coryphantha vivipara* (Plate K) and *Pediocactus simpsonii*, occur where wintertime temperatures commonly dip below −20°C. Two cacti, *Oroya peruviana* and *Tephrocactus floccosus*, grow at 4700 meters (15,400 feet) in the Andes of Peru. On the other hand, platyopuntias cultivated for fruit or for fodder can be harmed by occasional cold spells of −5°C to −10°C (14°F to 23°F). Although the process and mechanism of freezing in plants are active research areas with many unanswered questions, water clearly plays a key role in the freezing behavior of agaves and cacti.

To help understand how freezing temperatures can damage agaves and cacti, we will consider the events that occur in the chlorenchyma as the air temperature is lowered to −20°C (Figure 5.4). In response to the lowered air temperature, the tissue temperature steadily decreases to about −5°C, which is far lower than the freezing temperature expected based on the amount of dissolved molecules within the cells. Specifically, the dilute solution in the cells of agaves and cacti would freeze at −0.5°C to −1.0°C if given enough time. The observed decrease among plants below this *equilibrium* freezing temperature is a common phenomenon referred to as *supercooling*. Water maintained in a liquid state at subzero temperatures by means of supercooling can allow diffusion and metabolic processes to continue in the cells. It is only a temporary phenomenon, however, and is of little consequence to the tolerance of sustained low temperatures by agaves and cacti.

After supercooling leads to a tissue temperature of about −5°C, the tissue temperature rises (Figure 5.4), indicating that heat is being released in the tissue, a process called an *exotherm*. In this case, the exo-

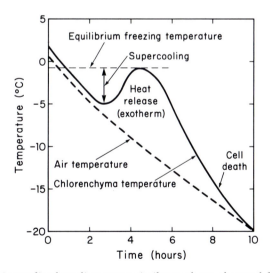

Figure 5.4   Generalized cooling curve similar to those observed for the chlorenchyma of *Carnegiea gigantea*, *Coryphantha vivipara*, and *Opuntia ficus-indica*. After supercooling below the equilibrium freezing temperature, the tissue temperature rises for a few hours. Continued decreases in chlorenchyma temperature are accompanied by death of the cells. Further details are provided in Nobel (1988).

therm is caused by the freezing of water, which releases the same amount of heat as is required to melt that mass of ice (known as the heat of fusion). The nuclei (starting locations) for the ice crystals are located on or near the outer surface of the cell walls surrounding the chlorenchyma cells; the water within the cells remains liquid. Water molecules then move from inside the cells and join the extracellular ice crystals, releasing heat. This keeps the chlorenchyma temperature considerably above that expected based on the initial part of the cooling curve (Figure 5.4). The amount of heat given off during the exotherm can be used to estimate how much water freezes extracellularly. When the tissue temperature again begins to decrease after a few hours (Figure 5.4), most of the water that had been inside the cells has moved out of the cells and has become part of the extracellular ice crystals.

As the air temperature continues to decrease (Figure 5.4), water continues to diffuse out of the cells and to become incorporated into the growing ice crystals. The cells thus become progressively dehydrated (similar to Figure 5.1B). This dehydration by the movement of intracellular water to the extracellular ice crystals causes injury to the cells that is apparently indistinguishable from the injury caused when prolonged drought leads to massive loss of water from the shoot. Beginning with a fully hydrated shoot of *Coryphantha vivipara* or *Ferocactus acanthodes*, the amount of intracellular water lost during lethal freezing is approximately the same as the amount of water lost during lethal drought, just over 90 percent. Irreversible damage and the eventual death of the cells following loss of intracellular water (Figure 5.4) apparently involve various cellular processes and various sites of damage, especially those related to membranes.

Most agaves and cacti can tolerate lower temperatures when the air temperature gradually decreases over a period of days or weeks, which is generally what happens during the autumn and early winter. The lowering of the temperature at which cell death occurs (Figure 5.4) is a widespread phenomenon among plants and is termed *low-temperature acclimation* or *hardening*. For agaves and cacti, lowering the air temperature by 10°C (18°F) over 2 weeks leads to a low-temperature hardening averaging 1.5°C. Species that tolerate the lowest temperatures generally have the greatest degree of low-temperature hardening. For instance, as air temperatures decrease by 10°C over a few weeks, the low-temperature hardening is at least 3°C for *Agave deserti*, *A. parryi*, and *A. utahensis* and at least 5°C for *Opuntia fragilis* and *O. humifusa*. The appreciable capacity for low-temperature hardening allows these species to occupy habitats that experience freezing temperatures for prolonged periods.

Although the phenomenon of low-temperature hardening is widespread among plants, various simultaneous changes obscure the mechanism. Many species of agaves and cacti, most noticeably platyopuntias, lose tissue water as winter approaches. Even in the laboratory under wet conditions, *Opuntia ficus-indica* and other platyopuntias can lose 10 to 25 percent of their cladode water if the air temperature is lowered by 20°C (36°F) over a few weeks. The loss of water causes the cladodes to become thinner and flaccid, and often the plants bend over. The resulting prostrate position may be advantageous in the field during the winter, allowing dehydrated shoots of *Opuntia erinacea, O. humifusa,* and *O. polyacantha* to lie under the snow, a position that is generally warmer than if the shoots were vertical above the snow. The morphological changes accompanying dehydration can consequently be crucial for the survival of these species during the winter.

Loss of shoot water during low-temperature hardening decreases the amount of water in the cells. Compared with the fully hydrated condition, less water in the cells leads to less loss of intracellular water during freezing episodes. Less water initially would also lead to smaller extracellular ice crystals, reducing the mechanical damage that such crystals could cause to the cells. In addition, less water might discourage the formation of intracellular ice crystals, which are essentially always lethal. Whatever the cellular basis, most gardeners recognize the importance of withholding water from agaves and cacti when periods of freezing temperatures approach, as hydrated shoots generally suffer more damage than do partially dehydrated ones during low-temperature episodes.

During acclimation to low temperatures, there also are major changes in the amount and the type of organic molecules in the stems of platyopuntias. The amount of the complex polysaccharide mucilage (Chapter 3) tends to increase within the extracellular spaces at low air temperatures. Mucilage may act as a nucleator for ice crystals, allowing more but smaller crystals to form in the extracellular region. The proliferation of many small ice crystals outside the cells could prevent the formation and penetration of single, large ice crystals into the cells. The amounts of various sugars, including glucose, fructose, and sucrose, also increase in the cells of agaves and cacti during acclimation to low temperature. For other plants, these sugars help protect various cellular constituents during exposure to freezing temperatures, although their role for agaves and cacti is unknown. It is clear that (1) the amount and the location of water play key roles in both freezing damage and low-temperature hardening; (2) solutes such as sugars affect the properties of water, especially at membrane surfaces within cells; and (3) mucilage and other extracellular molecules may also be important for the freezing process.

High Temperature

Most agaves and cacti grow in arid and semiarid regions that have extremely high summertime temperatures, but such temperatures are generally not lethal to mature plants. Indeed, mature agaves and cacti are among the most tolerant of high temperatures of all plant species. This is especially evident for various "dwarf" cacti like *Ariocarpus fissuratus, Epithelantha bokei,* and *Mammillaria lasiacantha* in the Chihuahuan Desert. Their small stature places their stems close to the soil surface, where temperatures in the summer can exceed 70°C (158°F) in a desert.

The seedlings of agaves and cacti are more vulnerable to tissue temperatures above 50°C than are large mature plants. These small young plants can also be more directly exposed to the high temperatures near the soil surface. A mature agave generally has a series of dead leaves in immediate contact with the soil, and the stems of most mature cacti also generally have no living chlorenchyma near the soil surface. The extremely high temperatures possible at the soil surface are consequently not as damaging to mature plants as they are to seedlings.

Most vascular plants succumb to high-temperature damage at 50°C to 55°C (122°F to 131°F), but nearly all agaves and cacti can tolerate such temperatures when properly acclimated. Agaves and cacti can undergo considerable high-temperature hardening, thereby aiding in their survival in deserts. As the average air temperature rises from 25°C (77°F) to 45°C (113°F) over a period of weeks, the average maximal temperature leading to the death of their chlorenchyma cells increases by 10°C (18°F). This almost incredible amount of high-temperature hardening helps account for the high temperatures that agaves and cacti can survive in the field. Most species tested can withstand 1 hour at 60°C (140°F); many can tolerate 65°C; and cladodes of *O. ficus-indica*, when properly acclimated, can survive 1 hour at 69°C (156°F)! Chloroplast function is severely impaired at temperatures well below the extremely high temperatures that the shoots can survive. Nevertheless, reactions in the chloroplasts of agaves and cacti tend to be less affected by high temperatures than is the case for other plants.

Nurse Plants

Because the seedlings of many agaves and cacti are vulnerable to the high temperatures near the soil surface, often they can survive only under the protective shade of other plants, referred to as *nurse plants* (Figure 5.5). Nurse plants are necessary for the survival of seedlings of *Agave deserti, Carnegiea gigantea, Coryphantha pallida, Echinocactus platyacanthus, Ferocactus histrix, Neobuxbaumia tetetzo, Opuntia leptocaulis,* and many other species. As the protected plants grow larger, they compete more and more intensely with the nurse plants that originally protected them. Competition by large columnar cacti like *C. gigantea* and *N. tetetzo* can eventually lead to the death of the nurse plants that originally protected them.

Perhaps the best-known cactus–nurse plant association is between

Figure 5.5   Seedling of *Agave deserti* sheltered by the nurse plant *Hilaria rigida*. The site is in the northwestern Sonoran Desert (Figure 1.5).

the saguaro (*C. gigantea*) and the paloverde tree (*Cercidium microphyllum*). Saguaro seeds germinate in response to summer rains, and their first survival test can be the subsequent drought. For those saguaro seedlings that survive their first drought, the most vulnerable time in southern Arizona is usually the hot period that occurs before the next summer's rains. In the northern half of Arizona or at high elevations, the most vulnerable time is during wintertime freezing episodes. Paloverde and other nurse plants are important for protecting saguaro seedlings from the extremes of both low and high temperature.

By intercepting sunlight, nurse plants can lower the maximal temperatures of the seedling tissues by more than 15°C (27°F). Lowering maximal summertime temperatures is probably the main way that nurse plants protect the seedlings of agaves and cacti, but they also interact in many other ways. The lower soil temperatures under the nurse plants reduce the rate of water evaporation near the soil surface, although water uptake by the nurse plant can eliminate this advantage to the seedling. Nurse plants can extend the distribution of saguaro into colder regions in Arizona and of *Trichocereus chilensis* into colder regions in Chile by sheltering seedlings from low-temperature extremes, just as the apical spines and pubescence do for barrel and columnar cacti. Thus the leaves and branches of a nurse plant act as an awning that protects a young seedling from the extremely low temperatures of a cold clear winter night in some regions and from the extremely high temperatures near the soil surface during a hot summer day in other regions. Nurse plants also hide tender young seedlings from browsing animals. In terms of the numbers of plants protected, this camouflage role of nurse plants can be more important than protection from low or high temperatures in those years with more moderate temperature extremes.

At a site in the northwestern Sonoran Desert (Figure 1.5), essentially all seedlings of *Agave deserti* surviving for 1 year or more are found in sheltered microhabitats, especially those provided by nurse plants (Figure 5.5). The most common nurse plant for *A. deserti* is the desert bunchgrass *Hilaria rigida*. Soil surface temperatures in the open can reach 71°C (160°F), which is lethal to the agave seedlings. Even the temperatures near the south side of the bunchgrass are too high, so seedlings become established only beneath the center or the northern part of the nurse plant. The photosynthetic photon flux (PPF) over the course of a day under this nurse plant averages about 65 percent lower than it is in an exposed location. The agave seedlings growing under nurse plants thus receive less light, which slows their growth.

During the first few years after germination, all the roots of the seedlings of *A. deserti* are in the upper 8 centimeters (3 inches) of the soil, where approximately half of the bunchgrass roots also occur. The agave seedlings are therefore subjected to intense competition for water, which further reduces their growth. On the other hand, the nutrient levels in the soil under this nurse plant are considerably higher than the average nutrient levels for exposed soil. The relatively extensive root system of the bunchgrass takes up nutrients from a wide region of the soil. When the leaves of the nurse plant die and fall off, their nutrients are returned to the soil near the agave seedling. This "mining" of nutrients from the surrounding soil raises the soil nitrogen level by 60 percent under the canopy of the bunchgrass, compared with that of an exposed location. The beneficial effects of increased nutrient availability under the nurse plant are offset by the diminished availability of light and water, causing the growth rate of agave seedlings to be about half as much as in an exposed location. This is the cost for survival, because without a nurse plant an agave seedling would almost surely die during its first summer.

REFERENCES

Fahn, A. 1990. *Plant Anatomy*. 4th ed. Pergamon Press, New York.

Franco, A. C., and P. S. Nobel. 1988. Interactions between seedlings of *Agave deserti* and the nurse plant *Hilaria rigida*. *Ecology* 69:1731–1740.

Goldstein, G., and P. S. Nobel. 1991. Changes in osmotic pressure and mucilage during low-temperature acclimation of *Opuntia ficus-indica*. *Plant Physiology* 97:954–961.

Loik, M. E., and P. S. Nobel. 1991. Water relations and mucopolysaccharide increases for a winter hardy cactus during acclimation to subzero temperatures. *Oecologia* 88:340–346.

Martin, J. T., and B. E. Juniper. 1970. *The Cuticles of Plants.* Edward Arnold, Edinburgh.

McAuliffe, J. R. 1984. Sahuaro–nurse tree associations in the Sonoran Desert: Competitive effects of sahuaros. *Oecologia* 64:319–321.

Nobel, P. S. 1980a. Influences of minimum stem temperatures on ranges of cacti in southwestern United States and central Chile. *Oecologia* 47:10–15.

Nobel, P. S. 1980b. Morphology, surface temperatures, and northern limits of columnar cacti in the Sonoran Desert. *Ecology* 61:1–7.

Nobel, P. S. 1982. Orientations of terminal cladodes of platyopuntias. *Botanical Gazette* 143:219–224.

Nobel, P. S. 1988. *Environmental Biology of Agaves and Cacti.* Cambridge University Press, New York.

Nobel, P. S., G. N. Geller, S. C. Kee, and A. D. Zimmerman. 1986. Temperatures and thermal tolerances for cacti exposed to high temperatures near the soil surface. *Plant, Cell and Environment* 9:279–287.

Steenbergh, W. F., and C. H. Lowe. 1977. *Ecology of the Saguaro: II. Reproduction, Germination, Establishment, Growth, and Survival of the Young Plant.* National Park Service, U.S. Government Printing Office, Washington, D.C.

Valiente-Banuet, A., A. Bolongaro-Crevenna, O. Briones, E. Ezcurra, M. Rosas, H. Nuñez, G. Barnard, and E. Vazquez. 1991. Spatial relationships between cacti and nurse shrubs in a semi-arid environment in central Mexico. *Journal of Vegetation Science* 2:15–20.

6

CO$_2$ Uptake by Plants

When Benjamin Heyne noticed in 1813 that a cactus tasted bitter in the early morning but lost its acidic taste by the late afternoon, he had stumbled on the foremost physiological adaptation of agaves and cacti (Chapter 1). Stomates of these plants open mainly at night, and the ensuing CO$_2$ uptake leads to a gradual acidification of the shoot—the CAM (Crassulacean acid metabolism) pathway. Gas exchange takes place when temperatures are lower than they are during the daytime, thereby greatly reducing water loss from the shoots of CAM plants. Agaves and cacti thus accumulate CO$_2$ during the night for daytime use in photosynthesis and simultaneously conserve the water stored in their shoots. We will now more closely consider the various physiological and anatomical underpinnings of this water-conserving method of taking up CO$_2$ for agaves and cacti compared with those of other plants.

The capture and retention of CO$_2$ for later use by plants require that it be bound to some molecule in the cells. Because this binding involves the rearrangement of chemical bonds, our first topic is the *biochemistry* (chemical reactions taking place in living organisms) of the three different pathways in plants that lead to *CO$_2$ fixation* (chemical binding of CO$_2$ to organic molecules). Although we will keep the details to a minimum, we will discuss both the cellular location of the reactions and their energetic costs. The overall patterns of gas exchange between plants and the atmosphere—CO$_2$ gains and losses as well as water vapor losses by the plant—are presented over 24-hour periods for measurements that can be made in the laboratory or the field (Figure 6.1). This leads us to the concept of water-use efficiency, a measure of the water cost for plants to take up CO$_2$ and hence to grow. Water-use efficiency is a useful standard for comparing the performance of the three biochemical pathways of photosynthesis.

Figure 6.1   Measurement of CO_2 uptake and water loss from leaves of *Agave mapisaga* in Tequexquinahuac, Mexico, near Mexico City. The plants are used for pulque production in the Valley of Mexico.

Binding of CO₂

As a gas, carbon dioxide occurs in only one form, CO_2. However, biochemical reactions take place in aqueous solutions where CO_2 can interact chemically with water. For example, a molecule of CO_2 can combine with a molecule of water (H_2O), leading to H_2CO_3, a molecule that can dissociate to form bicarbonate (HCO_3^-) and a hydrogen ion (H^+):

$$CO_2 + H_2O \rightleftharpoons H_2CO_3 \rightleftharpoons HCO_3^- + H^+ \qquad (6.1)$$

The symbol $\rightleftharpoons$ means that the reaction can proceed in either the forward direction or the reverse direction, depending on the relative concentrations of the molecules involved on the two sides of the double arrow.

The reactions in Equation 6.1 occur rapidly and do not require a *catalyst*, a substance that binds the reacting molecules and speeds up the reaction. Most reactions in living organisms require catalysts, which also add specificity, because only those molecules with the appropriate structure can be bound and enter into the reactions. The catalysts for biochemical reactions are proteins (polymers of amino acids), termed *enzymes*, that bind specific molecules and greatly speed up the reactions. Two enzymes that are important for CO_2 fixation in plants are (1) phosphoenolpyruvate carboxylase (PEPCase, for short), which binds HCO_3^-, and (2) ribulose-1,5-bisphosphate carboxylase/oxygenase (Rubisco, for short), which binds CO_2. Rubisco is by far the most abundant protein on the earth and is present in the chloroplasts of all plants.

Three Different Pathways

Our understanding of the biochemistry of photosynthesis advanced spectacularly during and immediately after World War II. The war effort was responsible for the development of radioactive compounds that can be used to trace biochemical pathways. For instance, CO_2 can be made radioactive, or *labeled*, with ^{14}C, an unstable isotope (molecular form) of carbon that can be readily detected when it decays (breaks down) and emits an electron.

When plants like spinach are exposed to $^{14}CO_2$ in the light, the labeled CO_2 can diffuse into the leaves and bind to some molecule during photosynthesis. This molecule then becomes radioactive and thus can be readily detected. For most plants, the first detectable compound incorporating the carbon of labeled CO_2 is 3-phosphoglycerate (Figure 6.2). This radioactive three-carbon compound is produced within a few seconds of exposure to $^{14}CO_2$ in plants known as C_3 *plants*. Eventually, the radioactive 3-phosphoglycerate leads to sugars and other compounds using energy derived from light in the process of photosynthesis.

About 93 percent of the 300,000 species of plants are C_3 plants and use the C_3 photosynthetic pathway for fixing carbon. The C_3 pathway is often called the "Calvin cycle" in honor of Melvin Calvin, an American biochemist at the University of California, Berkeley, who was involved in its elucidation. Because of the contributions of another American biochemist, Andrew Benson, the pathway is also known as the "Calvin–

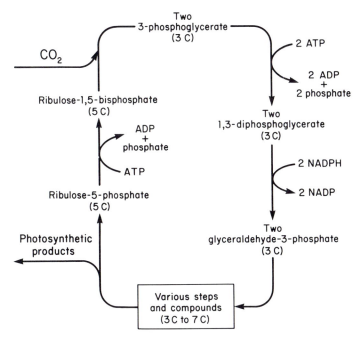

Figure 6.2   Relationship between certain compounds involved in the fixation of CO_2 into photosynthetic products by the C_3 pathway in chloroplasts. The overall cycle, which requires 3 ATP and 2 NADPH per CO_2, is known as the Calvin cycle or the photosynthetic carbon reduction cycle. The number of carbon atoms per molecule is indicated in parentheses beneath the compounds.

Benson cycle," as well as by the biochemically more descriptive "photosynthetic carbon reduction cycle" (Figure 6.2).

The experimental data for some plants did not fit the predictions of the Calvin cycle. Specifically, sometimes the four-carbon compounds malate and asparate were among the first radioactive compounds produced in the light. These products are formed in C_4 *plants* using the C_4 pathway. Only about 1 percent of all plants use the C_4 pathway, but they include agronomically important species such as corn (maize), sorghum, and sugarcane. The enzyme responsible for the initial binding of "CO_2" in the C_4 pathway is PEPCase, which catalyzes the following reaction:

$$HCO_3^- + PEP \leftrightharpoons oxaloacetate \qquad (6.2)$$

where PEP is phosphoenolpyruvate, a three-carbon compound, and oxaloacetate contains four carbon atoms. Oxaloacetate is converted within seconds to malate or aspartate.

The third pathway involved in CO_2 fixation in plants is the CAM pathway, which is used by about 6 percent of plant species. CAM species include the agronomically important pineapple, succulent plants in arid and semiarid areas, and tropical epiphytes that live nonparasitically on other plants, as do most orchids. This pathway, which includes aspects of both the C_3 and the C_4 pathways, is used by agaves and about 98 percent of cacti (all except leafy species in the genus *Pereskia* and a few other genera). The initial CO_2 fixation for CAM plants involves PEPCase and occurs at night, whereas CO_2 fixation for C_3 and C_4 plants occurs during the daytime. Plants exhibiting CAM also fix CO_2 during the daytime, but most of this CO_2 comes from compounds within the plants that had already incorporated CO_2 during the previous night. The day-time fixation of CO_2 in CAM plants uses the enzyme Rubisco and the C_3 photosynthetic pathway.

Cellular Location

Besides the differences in the enzyme that initially binds CO_2, the three photosynthetic pathways differ in the cellular location for that binding (Figure 6.3). For the C_3 pathway, CO_2 is initially bound to Rubisco in the chloroplasts, where chlorophyll and all the enzymes needed for photo-synthesis are found (Figure 6.2). Because a chloroplast is a subcellular compartment in chlorenchyma cells (Plate B), the C_3 pathway (Figure 6.3A) involves but a single compartment for all the processes from the initial CO_2 binding to the formation of photosynthetic products.

For the C_4 pathway, CO_2 is also initially bound in chlorenchyma cells, but the binding by PEPCase takes place outside the chloroplasts in a region known as the *cytosol* (Figure 6.3B). The four-carbon organic acids are not processed into photosynthetic products in chlorenchyma cells. Instead, the acids diffuse out of these cells to other chloroplast-containing *bundle sheath cells*, which surround the vascular tissue (Figure 6.3B). After entering the bundle sheath cells, the organic acids are *decarboxylated*, meaning that CO_2 is released from them. The released CO_2 is then bound to Rubisco in the chloroplasts of the bundle sheath cells, leading to the same enzymatic steps and photosynthetic products as for the C_3 pathway (Figure 6.2). The C_4 pathway thus uses a series of enzymatic steps, beginning with PEPCase in the cytosol of chloren-chyma cells, to deliver CO_2 to chloroplasts of the bundle sheath cells. Once in the bundle sheath cells, CO_2 is fixed into photosynthetic prod-ucts via the Calvin cycle (the photosynthetic carbon reduction cycle), as for C_3 plants.

CO_2 is taken up by CAM plants primarily at night, and the CO_2 taken up is initially bound by PEPCase in the cytosol of chlorenchyma cells (Figure 6.3C). The resulting malate or other four-carbon organic acid is transported into a large central vacuole. *Vacuoles* are sacs that often occupy at least 90 percent of the cell volume and generally contain

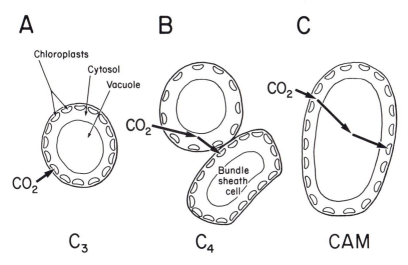

Figure 6.3 Cellular location for the initial binding and the subsequent fixation of CO_2 in plants from the three photosynthetic pathways: (A) C_3; (B) C_4; and (C) CAM. The three compartments involved are the chloroplasts, the cytosol, and the large central vacuole of chlorenchyma cells, as well as the chloroplasts of bundle sheath cells for the C_4 pathway (B). Processes take place during the daytime, except for the first two steps (arrows) for CAM plants.

only water and small solutes. Indeed, large vacuoles are a characteristic feature of the chlorenchyma cells of CAM plants (Figure 6.3C). The transport of malate into the vacuole during the night is an energy-requiring process that leads to the progressively more bitter, or acidic, taste of CAM plants during the night.

Beginning at dawn the following day, light stimulates the C_3 photosynthetic pathway (Figure 6.2) in the chloroplasts of CAM plants, just as it does for C_3 and C_4 plants. However, the major source of CO_2 for CAM plants during the daytime is from the decarboxylation of the malate and other organic acids that accumulated in the vacuoles during the previous night. Such stored malate diffuses out of the vacuoles during the daylight period (far right arrow, Figure 6.3C), and the CO_2 released internally by decarboxylation is bound by Rubisco in the chloroplasts, leading to photosynthetic products. As photosynthesis continues, more malate diffuses out of the vacuoles and becomes decarboxylated, thus supplying Rubisco with CO_2. Over a period of hours, the acidity of the chlorenchyma decreases and photosynthetic products accumulate. Because these processes do not require additional CO_2 from the atmosphere, the stomates of CAM plants can remain closed during the daytime. The higher daytime temperatures would lead to greater water loss through open stomates than would happen at night. Thus the nocturnal opening of stomates and their daytime closure is crucial for water conservation by CAM plants.

What happens when water conservation is not of overriding importance for CAM plants, as during periods of adequate rainfall? Except under conditions of drought, most CAM plants take up some CO_2 through open stomates during the daytime. This CO_2 acquired during the daytime is bound by Rubisco, just as is CO_2 taken up during the daytime by C_3 plants. Such CO_2 uptake and binding by CAM plants is often evident just after dawn, close to the coolest part of the day. Stomatal opening at that time does not lead to as much water loss as it would at noon or midafternoon, when higher tissue temperatures lead to greater water loss (Figure 1.9). Stomatal opening by some CAM plants also occurs in the late afternoon and is again accompanied by less water loss because temperatures are then lower than their maximal values. By late afternoon, all the malate accumulated during the previous night may have become decarboxylated and the CO_2 fixed by Rubisco, so photosynthesis would cease if the supply of CO_2 were not replenished. When water conservation is not of overriding importance, stomatal opening in the morning and in the late afternoon can lead to greater CO_2 uptake by CAM plants over 24 hours than with only nocturnal stomatal opening.

Energetics of CO_2 Fixation

Now that we have described the enzymes involved and the locations for the initial and subsequent binding of CO_2 by the three photosynthetic pathways, let us consider the energetic costs for such processes and related ones. The key is Rubisco, which can bind CO_2, leading to the incorporation of carbon into photosynthetic products (Figure 6.2). Unfortunately, Rubisco can also bind oxygen, O_2, which precludes the binding of CO_2.

The binding of O_2 to Rubisco leads to a series of reactions that release CO_2 and that are energetically expensive. The term *photorespiration* is used to describe this process—*photo-* because light is required, just as for photosynthesis, and *-respiration* because CO_2 is released, as is also the case for respiration. The various steps in the biochemical reactions of photosynthesis and photorespiration use energy carried by energy-rich compounds, often referred to as *energy currencies*. The two principal cellular energy currencies in biology are (1) *adenosine triphosphate* (ATP), which is made in chloroplasts in the light during photosynthesis and in other cellular compartments, *mitochondria*, continually during the day and night via respiration; and (2) an oxygen-poor or "reduced" compound, such as *nicotinamide adenine dinucleotide phosphate* (NADP, which in the reduced form is NADPH), which is also made in chloroplasts in the light.

Cellular Energy Currencies

Let us begin by comparing the "energy" of ATP and NADPH with the energy of the starting compounds used to form these two energy currencies. Such energetic considerations involve concentrations of various reactants and products in the biochemical reactions—concentrations that are hard to determine accurately in plant cells. Thus we generally use average cellular conditions to estimate how much energy is available when ATP and NADPH take part in biochemical reactions.

The energy released in biochemical reactions by such energy currencies is usually expressed per *mole* of ATP or NADPH reacting. A mole of a substance has a specific number of molecules, known as Avogadro's number (equal to 6.02×10^{23}). A mole of a substance also has a specific mass in grams—a mass that equals the sum of the atomic weights of the atoms making up its chemical structure. For instance, the atomic weight is 1 for hydrogen (H), 12 for carbon (C), and 16 for oxygen (O). A mole of water (H_2O) thus has a mass of 18 grams, and a mole of carbon dioxide (CO_2) has a mass of 44 grams. Energy is expressed in many different units. The unit most widely accepted for biochemical reactions is the *kilojoule*, in which the prefix *kilo-* means "times 1000" and a *joule* is the energy expended when a force of 1 newton acts over a distance of 1 meter. This latter definition, which is still rather abstract, can be illustrated using gravity. At sea level, the earth's gravity exerts a force of 9.8 newtons on 1 kilogram (2.20 pounds). The energy required to lift 1 kilogram vertically upward through a distance of 1 meter is 9.8 joules—the same energetic considerations apply when we climb steps or scale a mountain.

When ATP releases its stored energy, the molecule is split into two molecules, adenosine diphosphate (ADP) and a phosphate molecule. In addition, the components of water (H and OH) are incorporated at the site of bond breakage:

$$\text{ATP} + H_2O \rightleftharpoons \text{ADP} + \text{phosphate} \qquad (6.3)$$

where ATP and H_2O are the reactants and ADP and phosphate are the products.

The reaction that splits ATP—proceeding in the forward direction (to the right) in Equation 6.3—releases about 50 kilojoules of energy per mole of ATP split. In the reverse reaction, ATP is formed by the dehydration (removal of water) from ADP + phosphate, which occurs in chloroplasts and mitochondria. Because of the incredible number of water molecules present in cells, adding water to ATP can be relatively easy, but removing it from ADP + phosphate is difficult. Energetically, Equation 6.3 tends to proceed in the forward direction in aqueous solutions. ATP thus has higher energy than does ADP + phosphate, so ATP can act as an important energy currency in cells.

The use of NADPH as an energy currency involves transferring electrons and protons (H^+) between molecules. Just as the abundance of
water makes the energy stored in ATP readily available, the presence of
O_2 makes the reduced form, NADPH, a ready source of considerable
energy compared with the oxidized form, NADP. The formation of
NADPH involves removing electrons and protons from water, an energy-
requiring reaction that accompanies photosynthesis in chloroplasts. This
reaction releases oxygen, a crucial product of photosynthesis. The overall
process can be represented as

$$H_2O \rightleftharpoons {}^1\!/_2 O_2 + 2 \text{ electrons} + 2\ H^+ \qquad (6.4)$$

The electrons and protons released from water during photosynthesis
are exchanged by many compounds, eventually leading to the formation of NADPH from NADP.

The formation of the energy currency NADPH is usually represented
as

$$NADP + 2 \text{ electrons} + 2\ H^+ \rightleftharpoons NADPH + H^+ \qquad (6.5)$$

The net energy stored by the conversion of NADP to NADPH in Equation 6.5, when the electrons come from water (Equation 6.4), is 220
kilojoules per mole of NADPH. Because O_2 is present in essentially all
plant cells, NADPH tends to oxidize back to NADP, releasing much
energy and producing water. Specifically, O_2 binds the electrons and the
protons, which is a reversal of the reaction in Equation 6.4, without
performing any biologically useful work. Enzymes allow NADPH to be
an important energy currency by carefully transferring electrons and
protons to specifically targeted compounds in many biological pathways
(for example, Figure 6.2).

Photorespiration

Now that the two cellular energy currencies have been introduced, we
can evaluate the costs of the three photosynthetic pathways. For this, we
will consider photorespiration (Figure 6.4) as well as the extra steps of
the C_4 and the CAM pathways. Photorespiration is an apparently wasteful process that releases some of the CO_2 fixed by photosynthesis at a
high cost energetically. Rubisco can bind CO_2, leading to the incorporation of carbon into photosynthetic products, but it also can bind O_2
instead of CO_2. The binding of O_2 by Rubisco leads to a complex series
of reactions in the chloroplasts, in subcellular compartments known as
peroxisomes, and in the mitochondria (Figure 6.4). An average of 0.5
mole of CO_2 is released by these subsequent reactions per mole of O_2
bound to Rubisco. Various pathways are involved in the processing of
2-phosphoglycolate in photorespiration, pathways that consume approximately equal amounts of ATP and NADPH. These pathways have

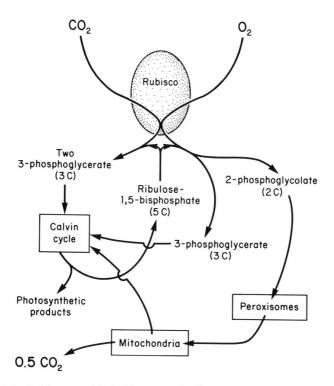

Figure 6.4 Rubisco can bind either CO_2, leading to photosynthesis, or O_2, leading to photorespiration. Reactions take place in the chloroplasts, except for the processing of 2-phosphoglycolate leading to the release of CO_2. Only a few of the compounds (the number of carbon atoms per molecule is indicated in parentheses) and a few of the pathways are presented.

an average energetic cost of 675 kilojoules per mole of O_2 bound to Rubisco.

Can a plant avoid the high cost of photorespiration? O_2 and CO_2 bind at apparently the same site on Rubisco, so raising the concentration of CO_2 relative to that of O_2 near Rubisco favors the CO_2 fixation pathway and reduces photorespiration (Figure 6.4). This is done artificially in many glasshouses (greenhouses) in which commercially important C_3 plants are grown. Specifically, burning gas, oil, coal, and even wood—often for the purpose of heating the glasshouse—raises the CO_2 level. The worldwide atmospheric CO_2 level in 1993 averaged 360 volumes of CO_2 per million volumes of air, generally referred to as 360 ppm (parts per million). Burning fossil fuels in commercial glasshouses raises their CO_2 levels to 800 to 1200 ppm. The result is an increase in CO_2 fixation and a decrease in photorespiration (Figure 6.4), which increases the growth of C_3 plants like chrysanthemums, roses, tomatoes, and strawberries that are raised in glasshouses.

CO_2 is initially bound by Rubisco in the chloroplasts of C_3 plants, whose photorespiration is important (Figure 6.4). For C_4 plants, CO_2 is initially bound by PEPCase in the cytosol, after which four-carbon organic acids are shuttled to the bundle sheath cells, where their decarboxylation greatly increases the CO_2 available for fixation by Rubisco (Figure 6.3B). This biochemical shuttle, which at first appears cumbersome because it involves extra steps and occurs between cells, thereby raises the CO_2 level high enough in the chloroplasts of the bundle sheath cells so that very little O_2 binds to Rubisco. Therefore, not much wasteful photorespiration occurs in C_4 plants. For CAM plants, CO_2 is also initially bound by PEPCase in the cytosol, but the process occurs at night and the organic acids are then actively transported into the vacuole of the same cell (Figure 6.3C). During the following daytime, the decarboxylation of such acids diffusing out of the vacuoles greatly raises the CO_2 level, so that CO_2 binding again dominates O_2 binding to Rubisco. Photorespiration is therefore also negligible for CAM plants, at least during the rapid decarboxylation of malate and other organic acids.

Energetic Costs

What are the respective energetic costs for the three photosynthetic pathways? Specifically, what is the cost in kilojoules for the net fixation of 1 mole of CO_2? The basic function of the Calvin cycle in the chloroplasts of plants from all three photosynthetic pathways is to incorporate carbon from CO_2 into a photosynthetic product. Each mole of CO_2 fixed via the Calvin cycle requires 3 moles of ATP and 2 moles of NADPH (Figure 6.2, Table 6.1). Based on the energy carried by these two energy currencies, the energy required for such CO_2 fixation in C_3 plants is

$$(3 \text{ moles ATP}) \times (50 \text{ kilojoules per mole}) + (2 \text{ moles} $$
$$\text{NADPH}) \times (220 \text{ kilojoules per mole}) = $$
$$590 \text{ kilojoules per mole of } CO_2 \text{ fixed} \qquad (6.6)$$

The extra steps for CO_2 fixation in C_4 and CAM plants entail additional energy costs compared with those for C_3 plants.

C_4 plants have three slightly different biochemical pathways for net CO_2 fixation. All three require 2 NADPH per CO_2 initially fixed by PEPCase, just as for C_3 plants. The other processes, including the biochemical shuttle from the chlorenchyma cells to the bundle sheath cells, require 1 or 2 more ATP per CO_2 than for C_3 plants. Hence, 4 or 5 moles of ATP and 2 moles of NADPH are required per mole of CO_2 fixed in C_4 plants, leading to an energetic cost of 640 or 690 kilojoules per mole of CO_2 initially bound (Table 6.1).

CAM plants have enzymatic and energetic requirements similar to those of C_4 plants, plus two additional costs. Specifically, 1 ATP is required to transport each malate into the vacuole at night (Figure 6.3C), and a polysaccharide must be formed during the daytime that will serve

Table 6.1   Energetics of CO$_2$ Fixation by C$_3$, C$_4$, and CAM Plants

	Pathways		
Quantity	*C$_3$*	*C$_4$*	*CAM*
Moles of ATP required per mole of CO$_2$ initially fixed	3	4 or 5	6 or 7
Moles of NADPH required per mole of CO$_2$ initially fixed	2	2	2
Energy cost in kilojoules per mole of CO$_2$ initially fixed	590	640 or 690	740 or 790
Mean energy cost in kilojoules per mole of net CO$_2$ fixation, including the cost of photorespiration*	867	665	765

Note: For details and references, see Nobel (1991b).

*Photorespiration was ignored for C$_4$ and CAM plants.

as the source for PEP (phosphoenolpyruvate; Equation 6.2) at night, with a net cost of 1 ATP per PEP. Thus the CAM pathway requires 2 more moles of ATP per mole of CO$_2$ initially fixed than does the C$_4$ pathway and so requires more energy, 740 or 790 kilojoules per mole of CO$_2$ fixed (Table 6.1). At this stage in our consideration of CO$_2$ uptake, the cost per mole of CO$_2$ initially fixed is least for the C$_3$ pathway and greatest for the CAM pathway.

The ranking among the three photosynthetic pathways for the cost of CO$_2$ fixation changes when photorespiration is included. For C$_3$ plants at moderate daytime temperatures, O$_2$ binds to Rubisco about one-quarter as often as does CO$_2$, leading to photorespiration instead of photosynthesis (Figure 6.4). Photorespiration becomes even more important above 30°C (86°F). For every 4 moles of CO$_2$ fixed by Rubisco into photosynthetic products, we will assume that 1 mole of O$_2$ binds to Rubisco, leading to the release of 0.5 mole of CO$_2$ (Figure 6.4). A net of only seven-eighths, or 0.875, as much CO$_2$ is therefore fixed, compared with the initially bound CO$_2$. The cost for this net CO$_2$ fixation per mole of CO$_2$ initially bound is increased by one-quarter of the cost of photorespiration, which we indicated costs 675 kilojoules per mole of O$_2$. Thus instead of 590 kilojoules per mole of CO$_2$ initially bound, the net cost for CO$_2$ fixation by C$_3$ plants is

$$\frac{(590 \text{ kilojoules per mole}) + (0.25 \times 675 \text{ kilojoules per mol})}{(0.875)}$$

$$= 867 \text{ kilojoules per mole of net CO}_2 \text{ fixation} \qquad (6.7)$$

The increase in the cost for net CO_2 fixation compared with initial CO_2 fixation for C_3 plants (Table 6.1) reflects the facts that some CO_2 is released by their photorespiration and that photorespiration is inherently an energetically expensive process.

We will assume that in C_4 and CAM plants the biochemical shuttles and the decarboxylation of organic acids raise the CO_2 level so high near Rubisco that photorespiration is negligible. Without the costs of photorespiration for C_4 and CAM plants, the cost per mole of net CO_2 fixation is lowest in C_4 species, intermediate for CAM species, and highest for C_3 species. Surprisingly, the most energetically expensive C_3 pathway (Table 6.1) is by far the most prevalent photosynthetic pathway among plants. We will reconsider this provocative conclusion in the next chapter, where we directly compare the productivities of plants from the three photosynthetic pathways under optimal growing conditions.

At this stage, we note that CAM plants are not disadvantaged compared with C_3 plants with regard to the biochemistry of net CO_2 fixation. The relatively slow growth of most CAM plants must then reflect the harsh environmental conditions under which many CAM plants occur naturally. In addition, other features more important than high rates of net CO_2 uptake and high rates of growth have evolved that ensure the survival of CAM plants in particular habitats. Yet some CAM plants can be highly productive, as demonstrated by the substantial harvests of various products from agaves and cacti as well as by the ability of certain platyopuntias to overrun particular ecosystems. Under certain simplifying assumptions, the energetic costs of net CO_2 fixation average about 13 percent higher for C_3 plants and 13 percent lower for C_4 plants compared with those for CAM plants (Table 6.1). Our discussion of the biochemistry of plants from the three pathways will help us in the next chapter to understand their relative productivities, and we will be able to appreciate how certain agaves and cacti can produce remarkably large amounts of biomass under favorable conditions.

Daily Patterns of CO_2 Exchange

Having discussed the energetic costs of the three photosynthetic pathways under optimal conditions, we next examine the maximal rates of net CO_2 uptake under such conditions (Figure 6.5). A consideration of CO_2 exchange over 24-hour periods for the leaves and stems of various plants in the field (Figure 6.1) inevitably glosses over cellular enzymatic details. Nevertheless, we will consider plants representing all three photosynthetic pathways.

The rate of CO_2 uptake is expressed as the amount of CO_2 (generally in moles) crossing the surface area of leaves or stems per time. For most purposes, the time unit is a second. The unit currently favored for surface area is 1 square meter (10.8 square feet). Just as when specifying

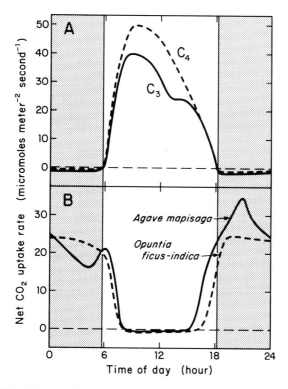

Figure 6.5   Net CO₂ uptake over 24 hours for various cultivated species under essentially optimal conditions of temperature, light, soil water status, and soil nutrients: (A) net CO₂ uptake rates averaged for six highly productive C₃ and C₄ crops; and (B) net CO₂ uptake rates for two highly productive CAM plants, *Agave mapisaga* and *Opuntia ficus-indica*. Data are on a projected area basis and are from Nobel (1988, 1991a, 1991b); Nobel, García-Moya, and Quero (1992); and unpublished observations. The stippled region represents night.

the area of a piece of cloth or a rug, we mean the area of one side, the area conventionally used for the thin leaves of C₃ and C₄ plants is that of one side—even though CO₂ may enter or exit from both sides, depending on the location of the stomates. The leaves of agaves and stems of cacti, however, are neither thin nor flat. The area of the upper surface of an agave leaf (Plate B) is less than the area of its lower surface, and many cacti have approximately round stems. To compare plants with different shoot morphologies—especially comparing C₃ and C₄ plants with thin leaves to agaves or cacti with thick leaves or thick stems—we will present data for CO₂ exchange on a projected area basis. Thus the total CO₂ uptake is divided by the area of one side of the leaf for a C₃ or a C₄ plant or by the area of the shadow cast by the broad side of an agave leaf, an opuntia cladode, or a barrel cactus stem. The modern conventional unit for CO_2 exchange is a micromole per meter squared per second, which we will represent by micromole meter^{-2} second^{-1}, where the prefix *micro-* means "times 10^{-6}" (0.000001).

The maximal rate of net CO_2 uptake for fast-growing, highly productive C_3 crops is about 40 micromoles meter^{-2} second^{-1} (Figure 6.5A). The rates of CO_2 uptake tend to be highest in midmorning and often dip during the hottest part of the day when the stomates partially close, which conserves water. The maximal rate of net CO_2 uptake for highly productive C_4 crops is somewhat higher than for C_3 crops, about 50 micromoles meter^{-2} second^{-1} (Figure 6.5A). This higher rate is consistent with the lower energetic costs for net CO_2 fixation because of the lack of photorespiration for C_4 plants (Table 6.1). As we would expect based on the biochemistry of C_3 and C_4 plants, there is no CO_2 uptake for either of them at night. Instead, there is then a small loss of CO_2 for C_3 and C_4 plants, because respiration in the mitochondria produces CO_2 that diffuses out of the leaves at night.

Unlike C_3 and C_4 plants, CAM plants take up most of their CO_2 at night (Figure 6.5B). Among agaves, the highest reported rate for net CO_2 uptake is 34 micromoles meter^{-2} second^{-1} for *Agave mapisaga* (Figure 6.1). Although not as high as the maximal net CO_2 uptake rates for C_3 or C_4 plants, such CO_2 uptake takes place at night and hence without concurrent photosynthesis. Moreover, *A. mapisaga* can take up CO_2 in the early morning (Figure 6.5B), presumably via the C_3 pathway, in which Rubisco directly binds the CO_2 diffusing into the leaves from the surrounding air. *Agave mapisaga* also takes up considerable net CO_2 in the late afternoon, again presumably using the C_3 pathway. Thus although the maximal rate of net CO_2 uptake is lower for *A. mapisaga*, the daily duration for CO_2 uptake is much longer than for C_3 or C_4 plants. *Opuntia ficus-indica* (Figure 1.7C) can take up 24 micromoles of CO_2 meter^{-2} second^{-1}, which is the highest reported rate of net CO_2 uptake among cacti, and it can sustain a high rate for most of the night. In addition, *O. ficus-indica*, like *A. mapisaga*, can use the C_3 pathway for considerable net CO_2 uptake during the cooler parts of the daytime in the morning and the late afternoon (Figure 6.5B).

These patterns of net CO_2 exchange are for cultivated crops under well-watered, optimal conditions. What are the patterns for agaves and cacti in native habitats under similar conditions? For two neighboring species in the northwestern Sonoran Desert, *Agave deserti* (Figure 1.5) and *Ferocactus acanthodes* (Figure 1.7A), net CO_2 uptake also occurs primarily at night (Figure 6.6A). However, the maximal rates of net CO_2 uptake are lower than those of cultivated CAM species, often about 10 to 14 micromoles meter^{-2} second^{-1}. The rates for these native species are more representative of net CO_2 exchange by CAM plants in their

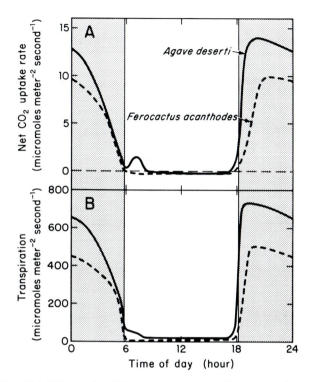

Figure 6.6   Net CO_2 uptake (A) and transpiration (B) over 24 hours for two neighboring species native to the Sonoran Desert, *Agave deserti* and *Ferocactus acanthodes*. Data are from Nobel (1988) for plants under essentially optimal conditions and are presented on a projected area basis (to facilitate comparison with C_3 and C_4 plants). The stippled region represents night.

natural habitats than are the rates for the agronomically important species *A. mapisaga* and *O. ficus-indica* (Figure 6.5B).

The lower maximal rates of net CO_2 uptake for the native species versus the cultivated species have a number of causes. *Agave deserti* and *F. acanthodes* have less PEPCase and Rubisco per shoot surface area than do *A. mapisaga* and *O. ficus-indica*. The biochemical reactions therefore cannot proceed as rapidly for the two native species. In addition, the pores of open stomates occupy a smaller fraction of the surface area for *A. deserti* than for *O. ficus-indica* (Figure 1.8B). It is therefore more difficult for CO_2 to diffuse into *A. deserti*, which is native to arid desert regions with limited rainfall, than into *O. ficus-indica*, which is cultivated, sometimes with irrigation, in semiarid regions and areas of moderate rainfall. Spines also screen some sunlight from the surface of *F. acanthodes*, whereas the surfaces of cultivated species are not appreciably shaded by spines.

The slower diffusion of CO_2 into the native species is correlated with a much slower rate of water loss than for the well-watered, cultivated species. Herein lies a trade-off, as high productivity is not always an advantage in an ecological context. Rather, water conservation, defenses against herbivores and pathogens, and successful reproduction can be the main forces for evolution. Under agricultural management, water and nutrients are often supplied, predators are generally controlled, and plants can be bred to be highly productive.

Water-Use Efficiency

Water loss by transpiration inevitably accompanies CO_2 uptake. For native CAM species in the field, stomatal opening and thus transpiration tend to occur at night (Figure 6.6B), which greatly reduces the rate of water loss compared with the same degree of stomatal opening during the daytime (Figure 1.9). This leads us to consider the water-use efficiency:

$$\text{Water-use efficiency} = \frac{\text{net moles of } CO_2 \text{ fixed}}{\text{moles of } H_2O \text{ transpired}} \qquad (6.8)$$

For CO_2 uptake to proceed at a substantial rate, the stomates must open, which for *Agave deserti* and *Ferocactus acanthodes* begins at dusk (Figure 6.6). Maximal CO_2 uptake generally coincides approximately with maximal transpiration for these species. Transpiration tends to decrease during the night for *A. deserti* and *F. acanthodes* (Figure 6.6B), which could represent stomatal closure restricting water loss. However, most of the decrease in transpiration during the night reflects a decrease in shoot temperature. As the shoot temperature decreases, the amount of water vapor per volume of air in the shoots become less—the amount approximately halves for each 10°C (18°F) decrease in temperature. The gradual decrease in shoot temperatures during the night thus reduces the tendency for water vapor to diffuse out. At dawn, the stomates of CAM plants may still be substantially open, and the internal level of CO_2 may be greatly lowered by the action of PEPCase. CO_2 can then diffuse in from the atmosphere and, in the presence of light, bind to Rubisco, leading to the pulse of C_3 photosynthesis observed at dawn for many agaves and cacti (Figures 6.5B and 6.6A). Because shoot temperatures are still low in the early morning, this pulse of CO_2 fixation is accompanied by low rates of transpiration.

The water cost of CO_2 fixation can be represented by the net CO_2 uptake over a 24-hour period divided by the accompanying amount of H_2O transpired (Equation 6.8). To determine this water-use efficiency, we sum the CO_2 exchange algebraically, counting CO_2 uptake as positive and CO_2 losses as negative. Essentially, we determine the area under the curves but above zero and subtract the area below the zero line

(Figures 6.5 and 6.6A). This leads us to the net moles of CO$_2$ taken up per meter^2 over the 24-hour period. Calculated as such, net CO$_2$ uptake is 0.43 mole meter^{-2} day^{-1} for *A. deserti* and 0.28 mole meter^{-2} day^{-1} for *F. acanthodes* (Table 6.2). Transpiration can also be summed (Figure 6.6B), and it is always positive, because water leaves a plant during both the daytime and the nighttime. The water loss is 26 moles meter^{-2} day^{-1} for *A. deserti* and 17 moles meter^{-2} day^{-1} for *F. acanthodes* (Table 6.2). To help appreciate the amount of water involved, we note that the daily losses are equivalent to water depths of 0.5 millimeter and 0.3 millimeter, respectively (1.0 millimeter = 0.04 inch). To obtain the water-use efficiency, the CO$_2$ uptake is divided by the water loss, both of which are expressed in moles meter^{-2} day^{-1}. For *A. deserti* and *F. acanthodes* in the field under conditions of wet soil, this water-use efficiency (Equation 6.8) averages 0.016 (Table 6.2).

What is the water-use efficiency for highly productive agricultural plants under well-watered conditions? Net CO$_2$ uptake over 24-hour periods (Figure 6.5) is relatively similar for such agronomic C$_3$, C$_4$, and CAM plants, ranging from 1.02 moles meter^{-2} day^{-1} for C$_3$ plants to 24 percent higher for C$_4$ plants (Table 6.2). However, CO$_2$ uptake by these C$_3$ and C$_4$ plants is accompanied by substantially greater amounts of water loss compared with that from CAM plants. Thus C$_3$ and C$_4$ plants have much lower water-use efficiencies than do CAM plants. For the cultivated CAM species *A. mapisaga* and *O. ficus-indica*, the amount of

Table 6.2 Net CO$_2$ Uptake, Water Loss, and Water-Use Efficiency for C$_3$, C$_4$, and CAM Plants

Species	Gas exchange over 24 hours (mole meter^{-2} day^{-1})		Water-use efficiency (CO$_2$/H$_2$O)
	Net CO$_2$ uptake	*Net water loss*	
Native			
Agave deserti	0.43	26	0.0165
Ferocactus acanthodes	0.28	17	0.0164
Cultivated			
C$_3$*	1.02	1130	0.0009
C$_4$*	1.26	740	0.0017
Agave mapisaga	1.17	230	0.0051
Opuntia ficus-indica	1.08	196	0.0055

Note: Data are based on totals over 24 hours under field conditions that were esentially optimal for CO$_2$ uptake and are from Figures 6.5 and 6.6 as well as literature cited in their captions.

*Values for C$_3$ and C$_4$ plants are based on the six most productive species using each pathway.

water transpired per square meter per day is three to five times lower than for the highly productive C_3 and C_4 crops. The water-use efficiency (Equation 6.8) for these two highly productive CAM plants averages 0.0053, which is six times higher than for comparable C_3 plants and three times higher than for such C_4 plants (Table 6.2).

As the soil dries, the stomates of CAM plants tend to remain closed for more of the daytime, and stomatal opening begins later in the night. This reduces transpiration in proportion to the amount of stomatal closure but has less effect on net CO_2 uptake, because the enzymatic processes are not as directly affected by stomatal closure as is the rate of water loss. Thus the water-use efficiency (Equation 6.8) tends to increase during the initial phases of drought. During prolonged drought, net CO_2 uptake decreases drastically and can become negligible. The water-use efficiency then also decreases. Because both CO_2 uptake and water loss can be reduced to very low values during prolonged drought, the water-use efficiency then has little meaning for plant performance.

The water-use efficiency is approximately three times higher for native CAM species than for cultivated CAM species under wet conditions. This disparity is caused by less stomatal opening at night for the native species as well as net CO_2 uptake during the daytime by the cultivated species. The daytime net CO_2 uptake, using the C_3 pathway with its lower water-use efficiency (Equation 6.8), leads to more CO_2 uptake per unit surface area but wastes more water compared with CO_2 uptake at night at lower temperatures. C_3 and C_4 plants cannot take up CO_2 at night, and hence they have inherently lower water-use efficiencies than do CAM plants (Table 6.2). When water conservation is a major management issue, the advantages of CAM species should be weighed, especially when the many current uses of agaves and cacti are evaluated economically.

REFERENCES

Edwards, G., and D. Walker. 1983. *C_3, C_4: Mechanisms, and Cellular and Environmental Regulation, of Photosynthesis*. University of California Press, Berkeley.

Nobel, P. S. 1988. *Environmental Biology of Agaves and Cacti*. Cambridge University Press, New York.

Nobel, P. S. 1991a. *Physicochemical and Environmental Plant Physiology*. Academic Press, San Diego.

Nobel, P. S. 1991b. Tansley Review No. 32. Achievable productivities of certain CAM plants: Basis for high values compared with C_3 and C_4 plants. *The New Phytologist* 119:183–205.

Nobel, P. S., E. García-Moya, and E. Quero. 1992. High annual productivity of certain agaves and cacti under cultivation. *Plant, Cell and Environment* 15:329–335.

Salisbury, F. B., and C. W. Ross. 1991. *Plant Physiology*. 4th ed. Wadsworth, Belmont, Calif.

Taiz, L., and E. Zeiger. 1991. *Plant Physiology*. Benjamin/Cummings, Redwood City, Calif.

7

Plant Productivity

Plant productivity is measured in various ways: the weight of agave fibers harvested, the number of boxes of a cactus fruit picked, or the total above-ground dry weight yield. Productivity reflects the cumulative effects of many factors on growth. For natural environments, soil water availability, temperature, photosynthetic photon flux (PPF), and soil nutrient content all can limit growth. Agricultural practices leading to high yields for economically important species often maximize growth by optimizing these factors. In deserts where the annual rainfall is generally less than 250 millimeters (10 inches), soil moisture is usually the key environmental factor limiting the growth of agaves and cacti. Agaves are cultivated for fiber or for beverage production in much wetter regions that have annual rainfalls of 600 to 1200 millimeters (24 to 47 inches). Certain cultivated cacti in arid and semiarid areas are even irrigated, as is the prickly pear *Opuntia ficus-indica* in central Chile. The productivity of a single plant is greatest when it is not shaded by other plants. On the other hand, productivity for an entire field is generally greatest when the plants are close together. This leads us to considerations of plant architecture, including shoot surface area per ground area, as well as the related PPF per shoot surface area.

To facilitate comparison among species, we will use the standard unit for productivity, the dry weight in tons (often spelled tonnes) per hectare per year, or tons hectare^{-1} year^{-1}. One metric ton is 1000 kilograms, and 1 hectare is 10,000 square meters (1 ton hectare^{-1} = 0.446 ton acre^{-1} for the slightly lighter "short" ton with 2000 pounds that is usually used in conjunction with acres). Because the roots of most crops are not used commercially and roots are difficult to harvest completely, biomass productivity generally refers to only the above-ground portions of the plants. The root:shoot ratio, which is about 0.1 for agaves and cacti (Table 4.2), can then be used to calculate the total biomass productivity. In this chapter we will show how environmental factors influence net CO_2 uptake per shoot surface area and how the spacing between plants can be chosen to maximize CO_2 uptake per ground area. The maximization of CO_2 uptake can enhance growth and productivity, which can be higher for certain agaves and cacti than for nearly all other cultivated crops, a remarkable conclusion whose cellular basis was outlined in the previous chapter.

Environmental Influences on CO_2 Uptake

We now turn to the effects of such environmental factors as soil water content, temperature, and PPF on the net CO_2 uptake of agaves and cacti. Our eventual goal is to relate CO_2 uptake to productivity for these CAM plants. We therefore must add up, or integrate, the net CO_2 exchange occurring during the day and the night, which equals the area under a curve for the instantaneous net CO_2 exchange over 24-hour periods (Figures 6.5 and 6.6). This daily net CO_2 uptake can be related to specific values for the environmental factors.

Water

Water uptake into roots occurs when the soil water potential (Ψ_{soil}) is higher (wetter) than the water potential of the root (Ψ_{root}; Chapter 4). After a few months of drought, Ψ_{root} for agaves and cacti may be -1.0 megapascal, whereas under conditions of wet soil, Ψ_{root} can be about -0.3 megapascal. After a major rainfall, Ψ_{soil} in the root region can increase to above -0.3 megapascal. In this situation, there is sustained water uptake, the stomates (Figure 1.8) can open fully, and net CO_2 uptake by the shoot can be maximal with regard to soil water status. As Ψ_{soil} decreases below Ψ_{root} and drought begins, there will be less stomatal opening and less CO_2 uptake. To quantify the influence of soil water availability on productivity, we need to consider the effect of drought on daily net CO_2 uptake.

Early in drought, the daily net CO_2 uptake of agaves and cacti is not noticeably affected (Figure 7.1). Stomatal opening and the biochemical reactions leading to CO_2 fixation are then fully supported by water stored in their succulent shoots. The period of drought during which daily net CO_2 uptake remains near maximal depends on the volume of water stored in the shoot relative to the surface area across which water can be lost by transpiration. This volume:area ratio is considerably higher for a thick-stemmed barrel cactus than for a thinner-leafed agave (Table 5.1). For *Agave deserti* (Figure 1.5) with a volume:area ratio of 0.6 centimeter, daily net CO_2 uptake remains near maximal for the first 4 days of drought (Figure 7.1). As the ratio increases from 1.1 centimeters for *Opuntia ficus-indica* (Figure 1.7C) to 6 centimeters for *Ferocactus acan-*

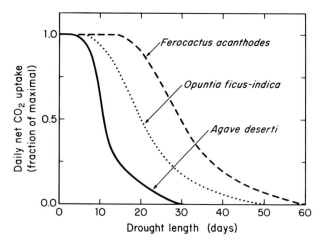

Figure 7.1 Influence of drought duration on daily net CO_2 uptake. Data represent the net CO_2 uptake integrated over 24 hours during drought relative to the maximal net CO_2 uptake integrated over 24 hours under conditions of wet soil. The volume:area ratio is 0.6 centimeter for *Agave deserti*, 1.1 centimeters for *Opuntia ficus-indica*, and 6 centimeters for *Ferocactus acanthodes*. Data are from Nobel (1977, 1988), Nobel and Hartsock (1984), and unpublished observations.

thodes (Figure 1.7A), the time that the daily net CO_2 uptake can remain maximal increases from 7 days to 15 days.

The barrel cactus *F. acanthodes* can rely on stored water to support its daily net CO_2 uptake for 2 months (Figure 7.1). Even after 30 days of drought, its net CO_2 uptake over 24 hours is still half of the maximal amount occurring under well-watered conditions. In comparison, daily net CO_2 uptake decreases to half maximal in 20 days for *O. ficus-indica* and in 11 days for *A. deserti*. For most CAM succulents, the maximal stomatal opening occurs later at night as drought continues. Once the response of net CO_2 uptake to drought has been determined, which is usually done in the laboratory under carefully controlled conditions (Figure 7.1), the soil water status can be related to the net CO_2 uptake possible in the field.

Temperature

Net CO_2 uptake is also influenced by temperature, which varies throughout the day and the night. To test the effects of temperature experimentally, plants can be grown in the laboratory with controlled air temperatures, usually with the daytime temperature about 10°C (18°F) higher than the nighttime temperature. After about 1 week, the various temperature-dependent processes acclimate to a particular set of day/night air temperatures, and the daily pattern of net CO_2 exchange becomes stable. The net CO_2 uptake rate is then measured and integrated over a 24-hour period to determine the daily net CO_2 uptake for that temperature regime (Figure 7.2). The day/night air temperatures can then be changed to another set of values and CO_2 uptake measured again.

For *Agave deserti, Ferocactus acanthodes, Opuntia ficus-indica,* and many other CAM plants, daily net CO_2 uptake is greatest for day/night air temperatures near 25°C/15°C (77°F/59°F; Figure 7.2). Because net CO_2 uptake occurs mainly at night for these CAM plants, nearly all the influence of temperature on daily net CO_2 uptake is that of the nighttime temperature. The optimal nighttime temperature of 15°C is relatively low for the suite of metabolic reactions involved in net CO_2 fixation. Agaves and cacti from warm tropical regions, such as epiphytic species growing in tree canopies, may have considerably higher optimal temperatures for net CO_2 uptake than do the three succulents considered here, a matter that warrants further research.

At a site in the northwestern Sonoran Desert where *A. deserti* and *F. acanthodes* are neighbors (Figure 1.5), the daily maximal/minimal air temperatures averaged over a month generally range from 15°C/5°C (59°F/41°F) in the winter to 35°C/25°C (95°F/77°F) in the summer. At the relatively low day/night air temperatures of 15°C/5°C, the daily net CO_2 uptake for these two species decreases by an average of only 24

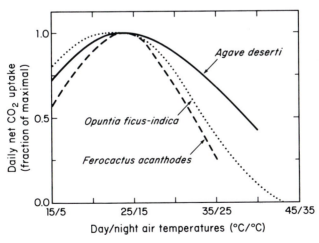

Figure 7.2 Influence of day/night air temperatures on daily net CO_2 uptake. Data represent the net CO_2 uptake integrated over 24 hours at particular day/night temperatures relative to the maximal net CO_2 uptake integrated over 24 hours under optimal temperatures. Data are from Nobel (1984, 1986) and Nobel and Hartsock (1984).

percent from its maximal values (Figure 7.2). At the moderately high day/night air temperatures of 35°C/25°C, the daily net CO_2 uptake for *A. deserti* and *F. acanthodes* is reduced an average of 52 percent. Compared with monthly variations in soil moisture, the average monthly temperatures have only a modest effect on the net CO_2 uptake—and hence growth—of these two species in this part of their native habitat. *Opuntia ficus-indica* can also have substantial net CO_2 uptake at quite low temperatures, as its daily net CO_2 uptake at 0°C (32°F) is 30 percent of the maximal value. *Opuntia humifusa* and *Tephrocactus floccocus* can have nocturnal net CO_2 uptake even at −5°C (23°F). Although various agaves and cacti can survive temperatures up to 65°C (149°F), high day/night air temperatures of 43°C/33°C (109°F/91°F) eliminate daily net CO_2 uptake by *O. ficus-indica* (Figure 7.2). Although low temperatures are more limiting for survival, high temperatures are generally more limiting for net CO_2 uptake by agaves and cacti in the field.

Photosynthetic Photon Flux

The standard way of presenting the photosynthetic light response for a C_3 or a C_4 plant is to relate the instantaneous photosynthetic photon flux (PPF) to the instantaneous net CO_2 uptake. Because the stomates of CAM plants tend to open at night, most of their net CO_2 uptake occurs in darkness, so this convention cannot be applied to them. Instead, the daytime availability of PPF can be related to the daily net CO_2 uptake for CAM plants. The integration of PPF over the daytime has some limitations for predicting CO_2 uptake ability—two days with the same daily (integrated) PPF, one with intermittent clouds and the other with a constant PPF, may not lead to exactly the same daily net CO_2 uptake. Day length can also affect daily net CO_2 uptake. Nevertheless, daily PPF has proved convenient for quantifying the light responses of net CO_2 uptake by agaves and cacti. The unit commonly used for daily PPF is moles of photons per square meter per day (moles meter^{-2} day^{-1}). Because individual photons, like molecules, can be counted, we refer to Avogadro's number (6.02×10^{23}) of photons as a mole of photons.

The responses of daily net CO_2 uptake to daily PPF are remarkably similar for *Agave deserti*, *Ferocactus acanthodes*, and *Opuntia ficus-indica* (Figure 7.3). At a very low daily PPF, below about 2 moles meter^{-2} day^{-1}, there is a daily net loss of CO_2. Some net CO_2 uptake may occur at night, but a greater net loss occurs during the daytime (for the slight loss of CO_2 that typically occurs near the middle of the day for CAM plants, see Figures 6.5B and 6.6). Parts of the shoot of an agave or a cactus exposed to a daily PPF below about 2 moles meter^{-2} day^{-1} thus do not contribute to the growth of the plant as a whole. Instead, the rest of the plant must supply the carbon that is lost from a part of the shoot whose daily net CO_2 uptake is negative. When the daily PPF increases to

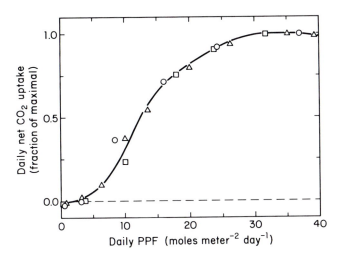

Figure 7.3 Influence of daily photosynthetic photon flux (PPF) on daily net CO_2 uptake. Data represent the net CO_2 uptake integrated over 24 hours at a particular daily PPF relative to the maximal net CO_2 uptake integrated over 24 hours for a saturating daily PPF of about 30 moles meter^{-2} day^{-1}. Data are for *Agave deserti* in the laboratory (O; Nobel, 1984), *Ferocactus acanthodes* in the field (Δ; Nobel, 1977), and *Opuntia ficus-indica* in the laboratory (□; Nobel and Hartsock, 1983).

about 12 moles meter^{-2} day^{-1}, the daily net CO_2 uptake by *A. deserti, F. acanthodes,* and *O. ficus-indica* becomes about half of the maximal value (Figure 7.3). As the PPF continues to increase and more CO_2 is refixed during photosynthesis, there is greater stomatal opening the following night to replenish the vacuolar malate that supplies the CO_2 for refixation. Daily net CO_2 uptake is maximal at a daily PPF of about 30 moles meter^{-2} day^{-1}.

How do these values of daily PPF compare with those available in nature? On a clear day at locations where the sun is approximately overhead at noon, the daily PPF on a horizontal surface is 60 to 70 moles meter^{-2} day^{-1}. But only seasonally, if ever, are the days clear with the sun approximately overhead at noon, and only a small fraction of the surfaces of agaves or cacti are horizontal. In fact, much of the surface of a cactus stem can be vertical. At nearly all latitudes and most times of the year, a vertical cladode intercepts the most daily PPF when facing east–west (Chapter 5). The maximal daily PPF for an east–west orientation is only 30 moles meter^{-2} day^{-1}. Because most surfaces of agaves and cacti do not receive a daily PPF of even 30 moles meter^{-2} day^{-1} on clear days, let alone on cloudy or overcast days, most of their surfaces are below PPF saturation for daily net CO_2 uptake (Figure 7.3). Thus even in the sunny environment of a desert, net CO_2 uptake and hence the growth of agaves and cacti are limited by the amount of light available.

High light, often in conjunction with relatively high temperature, can sometimes damage agaves and cacti. For example, if a houseplant acclimated to low indoor light is placed outside in the direct sun, the sudden increase in PPF can bleach the green leaves and stems, indicating a loss of chlorophyll. A "scorching," "sun scald," or "sunburn" can also occur in the field on the leaves of agaves and on the fruits and cladodes of platyopuntias. Exposure to direct sunlight shortly after sunrise can thus cause the death of certain surface tissues.

An Environmental Productivity Index

Daily net CO_2 uptake responds to the three main environmental factors of soil water status, temperature, and light (photosynthetic photon flux [PPF]; Figures 7.1, 7.2, and 7.3). Because all three factors can independently limit net CO_2 uptake, the overall limitation experienced by a plant can consist of simultaneous contributions from each factor. As a first approximation, we can consider that these limitations are multiplicative. For instance, when adequate soil water is available (Figure 7.1), the temperature may be so high that it halves the daily net CO_2 uptake (Figure 7.2). If the PPF is low and leads to half of the maximal daily CO_2 uptake (Figure 7.3), then the daily net CO_2 uptake may be only one-quarter of the value for wet conditions, optimal temperatures, and saturating PPF. If plants at these temperatures and PPF are droughted for a few months, their stomates will not open. Daily net CO_2 uptake is then zero, even though the temperature and the PPF are favorable for at least some CO_2 uptake.

The Basic Equation

The multiplicative nature of the effect of the three main environmental factors on daily net CO_2 uptake can be used to predict plant productivity. For this, an environmental productivity index, EPI, can be defined as

$$EPI = \text{Water Index} \times \text{Temperature Index} \times \text{PPF Index} \quad (7.1)$$

Each index ranges from 0.00 to 1.00 and indicates the fraction of maximal net CO_2 uptake possible under a given set of environmental conditions.

The Water Index equals 1.00 when the soil is wet and gradually decreases to 0.00 during drought (Figure 7.1). The Temperature Index for *Agave deserti*, *Ferocactus acanthodes*, and *Opuntia ficus-indica* is 1.00 for day/night air temperatures of about 25°C/15°C (77°F/59°F) and decreases for lower or higher temperatures (Figure 7.2). For these species, the PPF Index is 1.00 for a daily PPF of at least 30 moles meter^{-2} day^{-1}. As the PPF is lowered, the PPF Index steadily decreases and becomes 0.00 at about 2 moles meter^{-2} day^{-1} (Figure 7.3). Each of these three indices individually equals the fraction of maximal daily net CO_2 uptake. Their product, EPI (Equation 7.1), thus represents the fraction of maximal daily net CO_2 uptake for a particular environmental condition described by the individual indices.

EPI (Equation 7.1) represents only a first approximation for quantifying the influences of the environment on daily net CO_2 uptake as the three factors can interact. For instance, when the Water Index is reduced substantially by drought, the Temperature Index is maximal at a slightly lower temperature and the PPF Index is maximal at a slightly lower PPF. These factors interact when EPI is already substantially less than 1.00 and are of only secondary importance with regard to seasonal net CO_2 uptake. Nutrient levels in the soil can also substantially affect daily net CO_2 uptake and the growth of agaves and cacti. One way to describe the effect of soil elements is to determine the maximal daily net CO_2 uptake for plants growing in a specific soil when the three main environmental factors are not limiting. EPI times this maximal daily CO_2 uptake indicates the net uptake expected for a specific environmental condition. A more complex way is to analyze the soil for various key elements and to relate their levels to the daily net CO_2 uptake measured when an element is varied individually.

Five soil elements have been shown to have substantial effects on net CO_2 exchange and growth for several species of agaves and cacti. The macronutrients nitrogen, phosphorus, and potassium—which are the three main nutrients in commercial fertilizers—and the micronutrient boron have positive effects, whereas sodium has a negative effect (Table 4.3). The growth of agaves and cacti in sandy soil is maximal at approximately 3 grams of nitrogen, 0.06 gram of phosphorus, and 0.25 gram of potassium per kilogram dry weight of soil. Soil from the Chihuahuan Desert is deficient in boron, whereas soil from the Sonoran Desert has boron levels fully adequate for the growth of agaves and cacti (about 0.001 gram of boron per kilogram dry weight of soil). Lower levels of these four elements in the soil reduce daily net CO_2 uptake and growth accordingly. Salinity (sodium chloride) generally inhibits the growth of agaves and cacti, although the response depends on the species (Chapter 4).

Elevational Effects

Let us next consider a study that used the environmental productivity index (Equation 7.1) to predict the growth of *Agave deserti* over its entire elevational range in the northwestern Sonoran Desert (Figure 7.4). The study was undertaken to determine the effects of soil elements on growth, effects that were expected to vary considerably over the wide range in elevation. EPI was to be used to correct for differences in rainfall, temperature, and PPF among the sites. The differences in soil

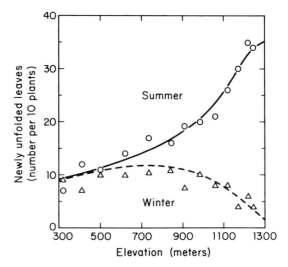

Figure 7.4   Variation in growth of *Agave deserti* measured by the unfolding of new leaves. The number of newly unfolded leaves was determined for a 3-month period in the summer (○) or in the winter (△). Leaf unfolding was also predicted (————, − − −) based on EPI from a previous year at a site of intermediate elevation in the northwestern Sonoran Desert (Figure 1.5). Data are adapted from Nobel and Hartsock (1986).

elements among the sites, however, did not appreciably affect growth, whereas predictions based on EPI closely matched the observed differences in growth.

The growth of agaves is easily monitored by counting the number of leaves that unfold from the central spike of folded leaves over a period of time (Figure 7.4), and the number of newly unfolded leaves can be related to various environmental factors for that period. In the summer, rainfall increases fourfold from the lowest site for *A. deserti* at an elevation of 320 meters (1050 feet) to the highest site at 1250 meters (4150 feet), which raises the Water Index by about 140 percent. The average temperature near midnight in the summer decreases from 27°C (81°F) at the lowest elevation site to 15°C (59°F) at the highest elevation site. Temperatures are thus more favorable for net CO_2 uptake (Figure 7.2) at the highest site, leading to a 70 percent higher Temperature Index there. The PPF Index does not vary appreciably with elevation in this case, so EPI (Equation 7.1) increases about fourfold in the summer with increasing elevation for the entire elevational range of *A. deserti*. This value closely agrees with the observed increases in growth as measured by leaf unfolding (Figure 7.4). In the winter, the increase in rainfall with elevation is more than offset, as far as leaf unfolding is concerned, by decreasing temperatures that become too cold for maximal CO_2 uptake. Predicted and observed growth is then least at the highest elevation (Figure 7.4).

EPI can help interpret differences in net CO_2 uptake and productivity due to simultaneous changes in various environmental factors, such as those accompanying changes in elevation. But EPI does not directly determine the survival or distributional limits of a species. For *A. deserti*, EPI on an annual basis is substantial at its lower elevational limit and is highest and thus most favorable for growth at the upper elevational limit of its natural occurrence. The lower elevational limit for this species is apparently caused by summertime temperatures that are too hot for seedlings to become established and the upper elevational limit is caused by competition for light with taller neighboring species.

Importance of Plant Spacing

Using the photosynthetic photon flux (PPF; Figure 7.3) and the environmental productivity index (EPI; Equation 7.1), we can quantitatively examine the effects of plant spacing on productivity. For this we need to know the maximal rate of net CO_2 uptake under ideal conditions, which when multiplied by EPI will give the actual net CO_2 uptake under particular field conditions. We also need to know the shoot surface area that intercepts PPF. The total leaf area per ground area for agaves is generally referred to as the *leaf area index,* and for cacti the total stem area per ground area is called the *stem area index.* The leaves of agaves and the stems of cacti are opaque three-dimensional objects whose sides face in different directions and hence receive different daily PPF. The different PPF levels and the total shoot area must be considered to calculate the net CO_2 uptake in the field. Because the daily net CO_2 uptake is expressed in moles per square meter of shoot surface area per day (moles meter^{-2} day^{-1}), the leaf or stem area index multiplied by this uptake will give the net CO_2 uptake per ground area per day.

To quantify the total daily PPF for a shoot is difficult, because the sun's position in the sky continually changes and the shoot surfaces of agaves and cacti face in different directions. Computer models, however, can calculate the angle between a particular part of the shoot and the position of the sun for specific latitudes, seasons, and times of day. Such models incorporate the effects of cloudiness, reflections of sunlight from the soil, and shading of one plant part by another for the complex shape of an agave rosette with many leaves or a platyopuntia with many cladodes. The shoot surface is divided into many parts, usually numbering in the thousands, and the PPF is added up hour by hour for each small surface. The integrated daily PPF for every part of the shoot surface is converted to daily net CO_2 uptake (Figure 7.3), which can be added up part by part to obtain the daily net CO_2 uptake for the entire plant. Such daily net CO_2 uptake can then be expressed per ground area and summed for an entire year.

To relate CO_2 fixation by photosynthesis to plant productivity, net CO_2 uptake is converted to dry weight gain. The main products of photosynthesis are sugars, which contain 1 mole of oxygen (atomic weight of 16) and 2 moles of hydrogen (atomic weight of 1) for each mole of carbon (atomic weight of 12). For each mole of carbon incorporated into a sugar, the dry weight of the plant increases by a mass in grams equal to the sum of these atomic weights (16 + 2 + 12), or 30 grams. Carbon represents a higher fraction of the dry weight of proteins, but nutrients from the soil also contribute some plant dry weight and contain no carbon. Therefore, a 30-gram gain in dry weight per mole of CO_2 taken up by the shoots is a reasonable estimate.

At low values of the leaf area index of an agave or the stem area index of a cactus, the plants are relatively far apart, with little mutual shading, and the annual productivity increases relatively linearly as the shoot surface area per ground area increases (Figure 7.5). If the plants are placed closer together, or when the leaves and stems grow and increase in surface area, shading occurs. This reduces the average PPF on any surface, lowers the daily net CO_2 uptake per shoot area (Figure 7.3), but increases the total CO_2 uptake per ground area. Productivity reaches a maximum at a leaf or a stem area index of 4 to 5 (Figure 7.5), for which essentially all of the PPF is intercepted by the plants.

As the leaf or the stem area index rises above about 5, annual productivity begins to decrease (Figure 7.5) when the reduction in daily CO_2 uptake per shoot area by mutual shading outweighs the gain in shoot area per ground area. As the leaf or stem area index increases, plants also incur a considerable cost for making and maintaining the ever-growing amount of nonphotosynthetic structural and water-storage tissue. The leaves of *Agave sisalana*, from which the fiber sisal is harvested, become structurally weak and have a low fiber content when the plants are so close together that considerable shading occurs. Plant productivity per unit ground area is therefore greatest at an intermediate leaf or stem area index, usually 4 to 5 for agaves and cacti. At such optimal spacings, the predicted productivity of henequen (*Agave fourcroydes*) is over 20 tons above-ground dry weight hectare^{-1} year^{-1} and that of the prickly pear *Opuntia ficus-indica* is nearly 40 tons hectare^{-1} year^{-1} (Figure 7.5). Compared with other plants, these are high biomass productivities, as we shall see in the next section.

Highest Productivities

In the previous chapter, we discussed the biochemistry of CO_2 fixation and how it varies among the three photosynthetic pathways: C_3, C_4, and

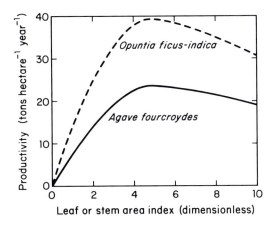

Figure 7.5   Predicted annual above-ground dry weight productivity for various shoot area indices. Environmental conditions for *Agave fourcroydes* are from Mérida, Yucatán, Mexico, and those for *Opuntia ficus-indica* are from Til Til, near Santiago, Chile. The PPF Index of EPI (Equation 7.1) varied with the spacing between the plants (Garcia de Cortázar, Acevedo, and Nobel, 1985; Nobel and Garcia de Cortázar, 1987). The carbon costs of the roots and the nonphotosynthetic parts of the shoot were assumed to equal 10 percent of the maximal daily net CO_2 uptake per shoot area and to be proportional to the plant biomass (Nobel, 1991).

CAM. The minimal energetic cost per net CO_2 fixed into photosynthetic products is highest for C_3 plants, lowest for C_4 plants, and intermediate for CAM plants like agaves and cacti (Table 6.1). We now return to a comparison of the three photosynthetic pathways, this time concentrating on maximal productivity.

CAM Plants

Agave deserti in the northwestern Sonoran Desert (Figure 1.5) had an annual above-ground dry weight productivity of 7 tons hectare^{-1} year^{-1} in a wet year with 430 millimeters (17 inches) of rainfall. In regions of Kenya and Tanzania with 1200 millimeters of rainfall, sisal (*Agave sisalana*) can yield 5 tons of dried leaf fiber hectare^{-1} year^{-1} and has a total above-ground dry weight productivity of about 20 tons hectare^{-1} year^{-1}. The latter yield is close to that predicted for henequen (*Agave fourcroydes)* under similar conditions (Figure 7.5). What is the maximal productivity for CAM plants when environmental factors are not limiting?

Computer simulations have been used to predict plant spacing and plant size that maximize the productivity of CAM species. For platyopuntias like *Opuntia amyclea* and *O. ficus-indica*, placing cladodes at 25-centimeter (10-inch) intervals in rows only 17 centimeters (7 inches) apart is predicted to be optimal for productivity. This density of 24 plants per square meter is similar to the 30 plants per square meter used to grow platyopuntias for nopalitos near Mexico City (Figure 3.3). A few months after the basal cladodes are planted, daughter cladodes develop, and the stem area index can increase to 4, which leads to maximal productivity for a platyopuntia (Figure 7.5). As growth continues, the plants can be pruned to maintain a stem area index of 4 to 5. Agaves can be planted close together so that very little PPF strikes the ground, PPF that is of little benefit for productivity (some of the PPF reflected from the soil surface, however, can be intercepted by plants and used for photosynthesis). These dense plantings maximize productivity per ground area. In practice, the open spaces generally needed for harvesting and other agricultural procedures cause the productivity per ground area for the entire field to be lower than such maximal values.

Besides closely spacing plants to maximize productivity with respect to PPF interception, productivity can be enhanced by providing sufficient soil water. The responses of agaves and cacti to soil water availability (Figure 7.1), as represented by the Water Index (Equation 7.1), indicate when productivity can be increased by irrigation. Without irrigation, the productivity of CAM species tends to be higher in regions with higher rainfall, especially if the rainfall is well distributed throughout the year. To achieve the highest productivities, the soil must also contain the necessary macronutrients and micronutrients or it must be fertilized.

When attention is paid to proper plant spacing and other management practices, the annual productivity of agaves and cacti can be extremely high (Table 7.1). For *Agave mapisaga* and *A. salmiana* in the

Table 7.1   Annual Productivities of the Most Productive CAM Plants

Species	Location	Productivity (tons hectare^{-1} year^{-1})
Agave mapisaga	Tequexquinahuac, Mexico, Mexico	38
Agave salmiana	Tequexquinahuac, Mexico, Mexico	42
Opuntia amyclea	Saltillo, Coahuila, Mexico	45
Opuntia ficus-indica	Santiago, Chile; Saltillo, Coahuila, Mexico; Milpa Alta, Federal District, Mexico	47 to 50
	Average	44

Note: Data are dry weights and are based on total above-ground harvests in years with substantial rainfall for the agaves and with irrigation for the cacti. Plants generally had a leaf or stem area index of 4 to 5. For details, see Garcia de Cortázar and Nobel (1992) and Nobel, García-Moya, and Quero (1992).

Valley of Mexico near Mexico City, the annual above-ground dry weight productivity can be 40 tons dry weight hectare^{-1} year^{-1}. For *Opuntia amyclea* and *O. ficus-indica* in Coahuila, Mexico, productivity can average 46 tons hectare^{-1} year^{-1}. *Opuntia ficus-indica* near Santiago, Chile, and in Milpa Alta, Mexico, can even produce 50 tons hectare^{-1} year^{-1} (Table 7.1)! These high productivities were achieved when plant spacing and soil water availability were close to optimal and the Temperature Index (Equation 7.1) averaged 0.84 to 0.90. Under the wet conditions provided, the dry weight was only 7 percent of the fresh weight for *O. ficus-indica*, so its fresh weight productivity was nearly 700 tons hectare^{-1} year^{-1}. Because these species have not been specially bred or carefully selected for maximal productivity, even higher productivities may be possible for agaves and cacti in the future.

C_3 and C_4 Plants

The high productivities of certain agaves and cacti (Table 7.1) can be better appreciated when the values are compared with the highest-producing C_3 and C_4 crops (Table 7.2). A principal goal for many agricultural species in the 1960s and the 1970s was to achieve higher annual productivities. Since then, breeding and selection have been used mainly to increase the yield of the specific plant part of economic interest in addition to preventing lodging for various grain crops, controlling pathogens, and improving tolerance of environmental stresses like salinity, drought, and low temperature.

The annual productivity for the four highest-producing agricultural species averages 36 tons hectare^{-1} year^{-1} for C_3 crops, 40 for C_3 trees, and 52 for C_4 crops (Table 7.2). The average productivity of the four highest-producing CAM species is 44 tons hectare^{-1} year^{-1} (Table 7.1), between that of the highest producing C_3 and C_4 species. This pattern of maximal productivity is consistent with the minimal energetic cost for net CO_2 fixation (Table 6.1). Yet the conclusion that certain agaves and cacti can have a higher productivity than the most productive C_3 plants, which make up 93 percent of all vascular species and dominate agriculture, is quite remarkable.

Very few agaves and cacti, however, have high productivities. Certainly a species of *Coryphantha* or *Mammillaria* only 10 centimeters (4 inches) tall when it is 50 years old could not be accused of high productivity. Species of short-statured cacti such as *Ariocarpus fissuratus* and *Epithelantha bokei* of the Chihuahuan Desert would not even be 10 centimeters tall at maturity. The many species of cacti whose stems are appreciably shaded by spines also do not have high rates of daily net CO_2 uptake or high productivities. Even if planted in monospecific stands and carefully managed, these species would produce far less than 10 tons hectare^{-1} year^{-1}, a typical productivity of well-managed C_3 and C_4 crops. Productivities also are low for epiphytic cacti growing in the

Table 7.2   Annual Productivities of the Most Highly Productive
C_3 and C_4 Plants

Type and species	*Location*	*Productivity (tons hectare^{-1} year^{-1})*
C_3 crops		
Beta vulgaris (sugar beet)	California, U.S.	30 to 34
Elaeis guineensis (oil palm)	Malaysia	40
Manihot esculenta (manihot)	Java; Sierra Leone	33 to 45
Medicago sativa (alfalfa)	Arizona and California, U.S.	30 to 34
	Average	36
C_3 trees		
Cryptomeria japonica	Japan	44
Eucalyptus globulus	Portugal	40
Eucalyptus grandis	South Africa	41
Pinus radiata (Monterey pine)	New Zealand	34 to 38
	Average	40
C_4 crops		
Pennisetum purpureum (napiergrass)	El Salvador; Puerto Rico	70
Saccharum officinarum (sugar-cane)	Guyana; Hawaii, U.S.; Queensland, Australia	50 to 67
Sorghum bicolor (sorghum)	California, U.S.	47
Zea mays (corn, maize)	Egypt; Italy; Peru; Colorado and California, U.S.	26 to 40
	Average	52

Note: Data are above-ground dry weights, plus below-ground harvests for *Beta vulgaris*. For references, see Nobel (1991).

canopies of tropical trees. Among agricultural CAM plants, probably the most widely known is the pineapple (*Ananas comosus*), which can have a maximal dry weight productivity in Hawaii of 30 tons hectare^{-1} year^{-1}. Clearly, CAM plants are not inherently low producers of biomass.

Water-use efficiency—which indicates the amount of carbon fixed into photosynthetic products per unit of water lost by transpiration (Equation 6.8)—also plays an important role in evaluating productivity for regions where soil water availability limits growth. Among cultivated species, CAM plants typically have a water-use efficiency about six times higher than that of C_3 plants and about three times higher than that of C_4 plants (Table 6.2). The higher water-use efficiency of CAM plants can be a distinct advantage for agaves and cacti in arid and semiarid regions with no irrigation. Even in wetter regions, soil moisture is often seasonally limiting, when the higher water-use efficiency of CAM plants can again be advantageous.

Interactions Among Organs

Some of the products resulting from net CO_2 uptake and photosynthesis are delivered to flowers, fruits, seeds, and other above-ground parts, and others are delivered to the roots. The delivery is by a vascular tissue known as the *phloem*, which has not been studied in detail for agaves or cacti. For other plant species, most photosynthetic products move in the phloem in the form of dissolved sugars, mainly sucrose.

Sucrose and other substances delivered to various organs in the phloem can be converted into compounds that become part of the dry weight of that organ. Alternatively, these substances can provide energy for many cellular processes. For instance, energy is needed for the active transport of nutrients from the soil into the root epidermal cells (Chapter 4) using ATP as the energy currency (Chapter 6). ATP is produced in chloroplasts in chlorenchyma cells during photosynthesis and in mitochondria in all cells and potentially at all times. The phloem delivers the sugars used by the mitochondria to produce ATP via *respiration*, which releases CO_2 through a series of chemical reactions that is basically the reverse of photosynthesis.

When an agave or a cactus is growing rapidly, most of the photosynthetic products that are delivered to the root system by the phloem are used for compounds that increase the root surface area and are thus incorporated into root dry weight. During prolonged drought, both shoots and roots stop growing. However, ATP is still needed to maintain the cells in a living state, and so sugars are required by the root cells to make ATP via respiration. Because the root:shoot ratio is quite low for agaves and cacti (Table 4.2), the amount of sugars required for this maintenance respiration in roots is relatively low. Low root demand for sugars also helps these plants tolerate extensive droughts without exhausting all their reserves produced by photosynthesis. When rainfall ends a drought, the first roots produced by agaves and cacti are generally fine lateral roots, which have a large surface area per dry weight. Thus water and nutrient uptake can increase greatly within a few days after rainfall with a minimal investment of biomass. Thicker roots that generally are initiated later when photosynthetic reserves are higher help expand the entire root system.

Photosynthetic products also move via the phloem to the reproductive structures of agaves and cacti, although some photosynthesis occurs in the outer green tissues of flowers and fruits. For *Agave deserti*, an enormous amount of dry weight is required to produce its massive flowering stalk, which is generally about 4 meters (13 feet) tall. In fact, the single reproductive effort of *A. deserti* requires the equivalent of the photosynthetic output of an entire year. The tremendous transfer of sugars and other organic compounds from the leaves of a mature agave to the rapidly growing flowering stalk contributes to the death of the plant after flowering. An enormous amount of water is also transferred from the leaves to the flowering stalk, about 18 kilograms (40 pounds). If the flowering stalk is cut off when relatively small, water and photosynthetic reserves are not transferred from the leaves, and *A. deserti* as well as other monocarpic agaves can continue to live for many years.

There is an interesting relationship between the dry weight of cladodes of *Opuntia ficus-indica* and their ability to support fruit growth (Figure 7.6). As the cladode surface area increases with growth, the dry weight of the cladode also tends to increase. But only those cladodes with a dry weight at least 33 grams (1.2 ounces) greater than the minimum value for a particular surface area produce fruit (Figure 7.6). Such an excess of dry weight over the minimum value is necessary, evidently as photosynthetic reserves, to supply the flowers and fruit with sugars and other organic compounds. If some of the flowers or young fruit are removed, the remaining fruit will become larger, as the photosynthetic reserves are then distributed among fewer organs. Such larger fruits are

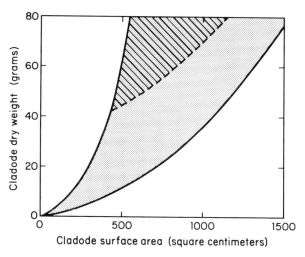

Figure 7.6 Relationship between total surface area (both sides) of cladodes of *Opuntia ficus-indica* and their dry weight (stippled area). Only those cladodes whose dry weight is at least 33 grams above the minimum value for a particular surface area (dashed line) support fruit growth (hatched area). One thousand square centimeters equals 155 square inches, and 60 grams equals 2.1 ounces. Data are from Garcia de Cortázar and Nobel (1992).

easier to harvest and also generally command higher prices per fresh weight. Shaded cladodes generally do not have much dry weight in excess of the minimum and thus generally do not produce fruit. Unshaded cladodes can have considerable dry weight in excess of the minimum and can produce up to 6 tons of fruit dry weight (30 tons of fresh weight) hectare^{-1} year^{-1}. For reasons that are so far unclear, such cladodes generally produce less fruit after their first bearing year. Much remains to be learned about the hormones and other factors that control the movement of photosynthetic products among organs via the phloem of agaves and cacti, information that could increase yields and benefit agricultural management in the future.

REFERENCES

Garcia de Cortázar, V., E. Acevedo, and P. S. Nobel. 1985. Modeling of PAR interception and productivity by *Opuntia ficus-indica*. *Agricultural and Forest Meteorology* 34:145–162.

Garcia de Cortázar, V., and P. S. Nobel. 1992. Biomass and fruit production for the prickly pear cactus, *Opuntia ficus-indica*. *Journal of the American Society for Horticultural Science* 117:558–562.

Nobel, P. S. 1977. Water relations and photosynthesis of a barrel cactus, *Ferocactus acanthodes*, in the Colorado Desert. *Oecologia* 27:117–133.

Nobel, P. S. 1984. Productivity of *Agave deserti*: Measurements by dry weight and monthly prediction using physiological responses to environmental parameters. *Oecologia* 64:1–7.

Nobel, P. S. 1986. Relation between monthly growth of *Ferocactus acanthodes* and an environmental productivity index. *American Journal of Botany* 73:541–547.

Nobel, P. S. 1988. *Environmental Biology of Agaves and Cacti*. Cambridge University Press, New York.

Nobel, P. S. 1989. A nutrient index quantifying productivity of agaves and cacti. *Journal of Applied Ecology* 26:635–645.

Nobel, P. S. 1991. Tansley Review No. 32. Achievable productivities of certain CAM plants: Basis for high values compared with C_3 and C_4 plants. *The New Phytologist* 119:183–205.

Nobel, P. S., and V. Garcia de Cortázar. 1987. Interception of photosynthetically active radiation and predicted productivity for *Agave* rosettes. *Photosynthetica* 21:261–272.

Nobel, P. S., E. García-Moya, and E. Quero. 1992. High annual productivity of certain agaves and cacti under cultivation. *Plant, Cell and Environment* 15:329–335.

Nobel, P. S., and T. L. Hartsock. 1983. Relationships between photosynthetically active radiation, nocturnal acid accumulation, and CO_2 uptake for a Crassulacean acid metabolism plant, *Opuntia ficus-indica*. *Plant Physiology* 71:71–75.

Nobel, P. S., and T. L. Hartsock. 1984. Physiological responses of *Opuntia ficus-indica* to growth temperature. *Physiologia Plantarum* 60:98–105.

Nobel, P. S., and T. L. Hartsock. 1986. Temperature, water, and PAR influences on predicted and measured productivity of *Agave deserti* at various elevations. *Oecologia* 68:181–185.

Nobel, P. S., B. Huang, and E. García-Moya. 1993. Root distribution, growth, respiration, and hydraulic conductivity for two highly productive agaves. *Journal of Experimental Botany* 44:747–750.

Pimienta, E. 1990. *El nopal tunero.* Universidad de Guadalajara, Jalisco, Mexico.

8

The Future

The physiology of agaves and cacti is remarkable in many ways. Certain species can have higher productivities than nearly all other plants (Tables 7.1 and 7.2). Their use of Crassulacean acid metabolism (CAM) ensures that water is conserved by nocturnal stomatal opening (Figure 1.9) without sacrificing the capacity for substantial net CO_2 uptake (Figure 6.5). CAM plants have a three- to sixfold higher water-use efficiency than do C_3 or C_4 plants (Table 6.2) and generally have a large water-storage capacity as well (Table 5.1). This can give agaves and cacti a competitive advantage over other plants for the one-third of the earth's land area that is arid or semiarid, as well as for the seasonally dry periods in other regions. The shallow roots of agaves and cacti (Figure 4.1) readily take up water from wet soil but do not lose much water to the extremely dry soils of deserts. Certain species can survive in regions without rainfall for many months or even years. We also cannot overlook the incredible temperature tolerance of agaves and cacti, ranging from $-40°C$ ($-40°F$) for *Opuntia fragilis* in Alberta, Canada, to $69°C$ ($156°F$) for *Opuntia ficus-indica* under field trials in California.

Equally remarkable are the many current uses for agaves and cacti, some of which date from prehistoric times (Chapters 2 and 3). Various Native Americans survived lean years eating virtually no other foods. In 1993, more than 1.2 million people worldwide depended on agaves and cacti for their primary source of income. Yet many products and uses of agaves and cacti are still largely unknown outside Latin America. The beverage pulque (Figures 1.1, 2.3, and 2.4; Plates D and E), the vegetable nopalitos (Figure 3.3), the cooked dish mixiote (Plate C), and the delicious pitaya fruit (Figure 3.2) are a few such examples. Information exchange, product development, and market research all are necessary to increase consumer awareness and acceptance of products from agaves and cacti in the twenty-first century. Institutions in Kenya and Tanzania that once studied agaves have been closed, and very few governments are now supporting coordinated basic research on agaves and cacti. Additional research, both physiological and agricultural, is greatly needed.

Our discussion of net CO_2 uptake by agaves and cacti (Chapter 6) and their resulting biomass accumulation (Chapter 7) included the influence of plant spacing on productivity (Figure 7.5). We developed the arguments step by step to help dispel the common myth of slow growth for agaves and cacti and to suggest new possibilities based on the extremely high productivities of certain species. The overrunning of the eastern coast of Australia by *Opuntia stricta* and related species in the twentieth century (Figure 3.9) attests to the competitive ability of various platyopuntias. An environmental productivity index (EPI; Equation 7.1) that quantifies the influences of environmental factors on net CO_2 uptake could have been used to predict the growth of these cacti in Australia. If the biological aspects of reproduction, consequences of spination, and vagaries of seed dispersal had also been appreciated, the invasive potential and ultimate ecological success of these plants might have been anticipated, thereby preventing folly and preserving fortunes.

Conservation

Conservation efforts for agaves and cacti are becoming more and more critical. The CITES convention (Chapter 1) is a welcome starting point for protecting threatened and endangered species, more of which occur in the family Cactaceae than in any other plant family. The beauty of collectable species of agaves and cacti can lead to the disappearance of local populations and sometimes to the extinction of entire species. Numerous cactus and succulent societies, environmental groups, and federal agencies support prudent ecological practices that should halt the unknowing loss of such species. Yet occasionally even knowledgeable and well-meaning enthusiasts say to themselves, "This one plant won't make a difference," and bring rare species from the wild into their collections.

Directors of botanical gardens, managers of large collections, and scientists are becoming increasingly aware of the importance of preserving *germplasm* (a general term referring to the genetic potential of a certain plant form). Considerable effort is now being expended in Chile, Italy, South Africa, the United States, and especially Mexico to establish germplasm "banks," or living collections, for various platyopuntias that are used for fruit or cladode production. Genetically diverse plants are collected from various regions and then maintained and propagated under controlled conditions. They can be reintroduced to the field from these banks to determine which species, varieties, or hybrids are well suited to local soil and climatic conditions.

Conservation endeavors often expose the confusing taxonomy of the genus *Opuntia*, which is economically the most important of the 122 genera in the Cactaceae. When the highly productive *Opuntia amyclea* (Table 7.1) is fertilized with pollen from another plant of *O. amyclea*, the mixed progeny are indistinguishable from *Opuntia ficus-indica, O. robusta*, and *O. streptacantha*. *Opuntia amyclea* is apparently a hybrid whose ancestry has become blurred over the years. Indeed, informal breeding and selection have improved the agronomic potential of *Opuntia*, but have created difficulties in identifying species and have complicated future breeding efforts to develop new varieties. Genetic diversity within a species is often a consequence of having multiple copies of the same chromosome in a cell, a condition known as *polyploidy*. Instead of two copies of the genetic information in a chromosome per cell, one from the male pollen and the other from the female part of the plant from which the seed develops (Figure 1.4), the cells of agaves and cacti typically have four, six, and even eight copies of a single chromosome. The mixing of chromosomes when reproductive cells divide leads to many possible combinations in polyploid species, resulting in dissimilar offspring.

Propagation techniques are crucial for threatened and endangered species as well as for agaves and cacti in general. Specimens do not have to be collected from the wild to satisfy the retail demand for more plants. Rather, cacti for the large ornamental market can be propagated by seed produced from plants that have already been collected and that are carefully pollinated (Figure 3.8). Most agaves and cacti can also be vegetatively propagated by cuttings. Agave plants produced as bulbils or ramets are important horticulturally as well as agriculturally for the fiber-producing henequen and sisal. Platyopuntias used for fruit, fodder, and forage are propagated almost exclusively by rooting detached cladodes. In addition, new techniques of propagation are now being developed. For instance, *Epiphyllum* raised for flowers (Plate L) can be readily grafted onto *Cereus* species and especially onto *Opuntia ficus-indica* (Figure 1.7C). To prepare the graft, the stems of *Epiphyllum* are trimmed to expose the vascular tissue, a small incision is made into an opuntia cladode, and the trimmed stem is inserted. In a few months, a dozen *Epiphyllum* stems can achieve more growth grafted onto a single cladode of the highly productive *O. ficus-indica* than the *Epiphyllum* could have achieved by itself in a few years! Interspecific and intergeneric grafts of other cacti have long been used commercially for the unusual and valuable forms that can result.

Each of the many areoles of a cactus can produce a new plant. This ability can be exploited by *tissue culture*, in which excised plant parts are provided with hormones and nutrients that ensure rapid growth (Figure 8.1). A little piece of tissue can be cut around the raised portion of an areole—a piece that includes the spines—for some endangered species and then placed into tissue culture. Cells in the meristem of the areole

Figure 8.1   Example of plantlets propagated from excised tissue of
Coryphantha robbinsorum, which is on the United States threatened species
list. The excised tissue was placed in agar containing hormones and nutrients,
leading to a fivefold increase in the number of shoots in one month.
(Photograph is courtesy of Gregory C. Phillips, Philip W. Clayton, and John F.
Hubstenberger.)

normally do not divide, because the apical meristem at the top of the
plant controls the cell division of the whole stem. But the apical meristem loses its dominance when the tissue is excised, and a bathing
solution containing growth hormones and nutrients generally ensures
much more rapid growth than would occur in nature (Figure 8.1).
Plants thus propagated from single areoles can be replanted in the wild,
thereby reversing the effects of overcollection and habitat destruction
that have caused many species of cacti to become threatened or endangered.

Propagation from individual cactus areoles can produce a few hun-
dred new plants from one parent plant (Figure 8.1), but micropropagation techniques for agaves and cacti offer much greater yields of new
plants. If the cells in a meristem can be teased apart, each cell can give
rise to an entire new plant. Cells in other plant tissues can also sometimes divide, especially under the favorable conditions of tissue culture
with the appropriate growth regulators. Thousands of plants can thus be
produced from tissue removed from a single desirable parent plant. A
whole range of modern biotechnological procedures is also possible

when working with isolated cells. One goal is to improve the germplasm by introducing new or modified genetic material. Such laboratory techniques may, for instance, greatly reduce the 8- to 10-year maturation time for the agaves used for fiber or beverage production.

Agaves

Early in the twentieth century, most of the cultivated area for agaves was devoted to species used for fiber, mainly henequen (*Agave fourcroydes*; Figure 2.6) in Mexico and sisal (*Agave sisalana*) in eastern Africa. The strong and visually attractive fiber from agave leaves has been made into baling twine, rope, sacks, and various fabrics used in crafts and interior decorating. When combined with the proper binders in composite substances, agave fibers can be used in the manufacture of flame-resistant structural materials with excellent tensile strength and durability. These construction materials, which can also have good insulating properties, are suitable for roofing, walls, and many other fiberboard applications. Agave fibers can also be incorporated into paper, which has been used for products ranging from crude cement bags to high-quality paper for currency.

The agave leaf–fiber industry has generally been slow to respond to changes in the marketplace. Since World War II, strong waterproof synthetic fibers have replaced natural agave fibers in most of their former roles. Yet secondary products from agave leaves could restore the profitability of raising henequen and sisal. Fibers of these species constitute only about one-quarter of their leaf dry weight; the remainder is generally discarded or used as fertilizer. Leaf pulp retained after the fibers are extracted is nutritious for cattle and alone could almost justify growing certain agaves. Using leaf pulp from henequen and sisal as a secondary product to feed cattle, sheep, and goats would improve the economics of the fiber industry. Although the leaves of fiber agaves are currently used as fodder to a limited extent, new field trials are needed to examine possibilities for species like *Agave mapisaga* and *A. salmiana* with even higher productivity (Table 7.1).

Sapogenins (Figure 2.7A) extracted from agave leaves could revolutionize agave cultivation. Sapogenins can be an important secondary product of the agave fiber and beverage industries and can be produced at a relatively low cost. But the real profitability may come from other species, such as *Agave vilmoriniana*, the present sapogenin-producing champion. The leaves of five other agave species are known to contain over 2 percent sapogenins by dry weight. Because 6 percent of the world's corticosteroids (Figure 2.7) are currently derived from sapogenins in agave leaves, a market base for increased commercialization already exists. An agave sapogenin that can be chemically modified or even a crude leaf extract from some agave may promote the growth of livestock. Human medicine may also benefit from extracts of agave leaves, which are currently touted by practitioners of folk medicine. Such extracts are used as anticolic, anti-inflammatory, and antispasmodic agents; as diuretics; and as aids in the relief of rheumatism. Future experimentation is needed to develop the full medicinal and agricultural potential of extracts from agave leaves.

The financial importance of the many beverages produced from agaves will undoubtedly increase in the future. Aguamiel is tasty, and so techniques should be developed to prevent its souring after collection. A more widespread distribution network is needed for pure or diluted aguamiel. Pulque, especially when flavored with extracts of various fruits, is usually enjoyed by first-time consumers. In 1993, the worldwide demand for tequila exceeded the production possible from agaves (Figure 2.5), and so other sources of carbohydrates or alcohol were used to increase the amount of tequila bottled. Mescal is also gaining in international popularity. Clearly, the many beverages made from agaves have a great potential for increased production.

Expanding the beverage market for agave products presents a real challenge to physiologists, biotechnologists, and plant breeders. The monetary gain would be much greater if the species used for tequila and mescal matured earlier. A farmer who can get a cash crop the first year after planting maize (corn, *Zea mays*) or the second year for sugarcane (*Saccharum officinarum*) must wait 8 or more years before the agaves reach maturity and can be harvested for alcohol production (Figures 1.1, 2.3, and 2.5). The markets are currently expanding for tequila and mescal, so breeding and selection efforts that shorten the time to maturity could pay large financial dividends.

Cacti

The world market for cactus fruits is already large (Table 3.1), but entrepreneurs in many countries are seeking ways to expand it. When people try cactus pears (usually *Opuntia ficus-indica*; Plate H) or pitayas (Figure 3.2) for the first time, adults and especially children are usually pleasantly surprised by their taste. In fact, the problem of acceptance is not taste but, rather, the nasty glochids on the fruit surface and the many seeds. Nearly all the glochids can be removed by brushing the fruit surface before marketing. To avoid glochids, the pulp can be scooped out of a fruit without gripping it with bare fingers. Breeding can reduce the number and the size of seeds, and seedless varieties may someday be developed. The seeds of the delicious pitayas, species that are generally in the genus *Stenocereus*, are smaller and more easily swallowed than are the seeds of cactus pears. Cactus seeds, however, readily pass through the digestive tracts of humans and can provide beneficial roughage. The high fructose content of cactus pears is also being promoted for those with diabetes.

A problem with marketing cactus pears is that the fruits are generally not available year-round. The practice of scozzolatura (Figure 3.1) can delay the time of fruit harvest. Judicious pruning or thinning based on the excess dry weight of the cladodes (Figure 7.6), as well as drip irrigation, can also modify the timing of fruit harvest. Developing varieties that mature at different times of the year and planting in a series of areas with different climates should help. Coordinating international marketing efforts for cactus pears in the Northern and Southern hemispheres—regions with a 6-month seasonal offset—can also contribute to year-round fruit availability.

Fruits of more than 40 species of cacti can be purchased in Latin America, although only a few species are cultivated on a major scale. The market for pitayas is increasing, so various species in the genera *Hylocereus* and *Stenocereus* are being cultivated in Colombia, Israel, Mexico, and other countries. *Opuntia fulgida*, which is native to the Sonoran Desert, has fruits that are tart but sweet and can be grown without irrigation in arid regions. Even the fruits of some *Mammillaria* species are tasty; such species could be cultivated and the small fruit picked like strawberries. In addition to fresh fruit, processed fruit can be used to prepare miel de tuna, queso de tuna, and other products popular in Mexico. Oil can also be obtained from the seeds of many species. The present uses of fruits of various cactus species is large but could be greatly expanded.

The use of young cladodes (nopalitos) as vegetables (Figure 3.3) has been minimal outside Mexico. This lack of popularity is due in part to the unfamiliarity with preparing nopalitos and the absence of fresh material year-round. The shelf life of nopalitos can be increased to about 30 days by refrigeration. More widespread information on their uses is currently needed—nopalitos are not only nutritious but also delicious, especially when marinated or cooked with onions and peppers. At an international meeting on opuntias in Mexico in 1992, over 60 dishes were prepared in a cactus-cooking contest, and nearly all of them included nopalitos. Recipes number in the hundreds and include the holiday favorite, mixiote (Plate C). Bottled marinated nopalitos are available in supermarkets and delicatessens in various countries, and fresh young cladodes are increasingly found alongside other fresh produce. Nopalitos are recommended for those with diabetes, high cholesterol, and prostate problems, and their worldwide consumption will most likely substantially increase in the future.

Mucilage may also find more uses. This viscous polysaccharide can constitute over one-third of the cladode dry weight for some platyopuntias. Industrial uses include adhesives, fiberboard binders, and a source for certain unusual sugars. Mucilage can be used as a thickener for soups, ice cream, and marmalades, similar to current uses of agar, gelatin, and other polysaccharides. Pharmaceutical applications and uses in herbal medicine have also been considered for cactus mucilage and may expand in the future.

In terms of the current land areas utilized and their future potential, the greatest use of platyopuntias may be for cattle fodder and forage (Figure 3.4), speculation based on the present demand for beef and other meats for human consumption. In 1992, more than 600,000 hectares of *Opuntia ficus-indica* and similar species were cultivated for fodder worldwide (Table 3.1). Cattle readily eat cladodes, and cladode productivity can be extremely high in terms of both dry weight (Table 7.1) and fresh weight. Increased research and breeding efforts are needed to identify and select species with a greater tolerance of freezing temperatures and salinity, as well as those with higher levels of nitrogen and phosphorus in their cladodes. Recently, a variety of *Opuntia stricta*, the past scourge of Australia, has been identified that meets these two nutritional needs of cattle.

Singeing spines off prickly pear cacti (Plate J) dates back more than 100 years in Texas. After singeing the cladodes of both naturally occurring and planted platyopuntias, their water and nutrients become accessible to cattle. Another interesting feed for penned cattle consists of roughly equal parts of chopped prickly pear, alfalfa, and chicken manure (Figure 8.2)! This nutritionally well-balanced, if somewhat repulsive, feed has been used to fatten cattle in various countries, with the cactus supplying moisture, sugars, and many of the needed nutrients.

Because of the high biomass productivity of certain agaves and cacti (Table 7.1), they should be used to generate energy, such as sources of biogas or alcohol for "gasohol." Small pilot projects have been initiated in Chile using *Opuntia ficus-indica* as the principal starting material for producing biogas. Chopped cladodes are placed in relatively small containers together with manure and household garbage; water is added; and the container is sealed from the outside air. Chemical reactions then take place that deplete the oxygen and lead to the production of methane, CH_4. Methane can be used for household cooking and heating as well as to power generators that produce electricity. The sludge remaining after the biogas is produced can be fed to cattle—it is somewhat similar to the prickly pear–alfalfa–chicken manure diet (Figure 8.2)—or it can be used as fertilizer. One of the challenges to make such biogas systems productive year-round is to devise strategies for periodically harvesting the cladodes of *O. ficus-indica* while retaining a stem area index that keeps biomass productivity near maximal (Figure 7.5).

Figure 8.2   Cattle eating chopped *Opuntia ficus-indica* mixed with approximately equal amounts of alfalfa and chicken manure. The penned animals are being fattened in Lampa, near Santiago, Chile.

Global Climate Change

Although the exact environmental consequences of the rising level of CO_2 in the atmosphere are unclear, this change is occurring over a geologically short period of time. The higher levels of CO_2 and other "greenhouse" gases, such as methane, may cause air temperatures to rise a few degrees Celsius by the middle of the twenty-first century, compared with mean values in the middle of the twentieth century (2°C to 4°C = 4°F to 7°F). The predicted changes are far more complicated than just increases in temperature. Large-scale shifts in air movement will occur, changing precipitation patterns. The effects of the resulting changes in soil water content, temperature, and light on daily net CO_2 uptake can be predicted for agaves and cacti under current CO_2 levels (Figures 7.1 through 7.3). Whether elevated atmospheric CO_2 levels will change the responses of these plants is not yet known.

The cellular location for the initial binding of CO_2 and its subsequent processing vary among the three photosynthetic pathways (Figure 6.3). The high levels of CO_2 produced in C_4 and CAM plants during the daytime overwhelm the oxygenase activity of Rubisco (Figure 6.4), so their photorespiration is very low. Elevated CO_2 levels have no direct major consequences for net CO_2 uptake under such conditions. On the other hand, the predicted doubling of atmospheric CO_2 levels should reduce the photorespiration of C_3 plants and thereby enhance their productivity. The productivity advantage of certain CAM plants over C_3 plants (Table 6.1) would thus decrease as atmospheric CO_2 levels rise. However, the appreciable net CO_2 uptake during the daytime for the most productive agaves and cacti (Figure 6.5) directly uses the C_3 pathway. This part of the net CO_2 uptake by CAM plants would receive the same benefit from elevated CO_2 levels that C_3 plants would receive. Of potentially greater importance for net CO_2 uptake and growth is the predicted change in rainfall patterns accompanying global climate change. The central part of the United States is predicted to become drier, which would favor CAM plants with their higher water-use efficiency (Table 6.2). The greater CO_2 fixation per water transpired by agaves and cacti than by corn and wheat might cause the present-day "breadbasket" of the midwestern states to become the "prickly pear" basket of the future! At the least, platyopuntias could be used more extensively to provide cladodes for cattle fodder as well as for fruit and vegetables.

The high potential productivity of certain agaves and cacti might even be used to help slow the trend of annually increasing atmospheric CO_2 levels. Such species could be planted in arid and semiarid regions that currently are not appreciably used agronomically, although any widespread usage of such CAM plants would involve major political and land-use decisions at national and even international levels. Nonetheless, the scientific basis for predicting the consequences of such plantings already exists.

Predictions

The past and present clearly indicate a bright future for agaves and cacti. We need only consider the myriad uses of agaves and cacti by Seri Indians (Table 1.1) as well as their current uses throughout Latin America and the rest of the world. The beauty and hardiness of these plants are appreciated more and more. Cacti can usually be purchased wherever household plants are sold. Many botanical gardens feature the dramatic shapes and glorious flowers of cacti (Plates K and L) as well as the radial symmetry and beauty of agaves (Plates F and G). The preservation of collected plants for enjoyment by future generations is seemingly assured, especially when the many unusual and bizarre forms of agaves and cacti are included (Figure 8.3).

The aesthetic appreciation of agaves and cacti is important for their

Figure 8.3   The author holding a bizarre form of *Opuntia ficus-indica*. The formation and fusion of about five cladodes during the early stage of development was apparently caused by a mycoplasma that disrupts the normal process of development in a meristem.

preservation, but the future of many species resides in their economic usefulness. A starting point is their potentially high productivity (Table 7.1). With careful breeding, selection, and biotechnological manipulations, astonishing advances can be made. The great morphological diversity among agaves and cacti, even in plants of supposedly the same species, favors such genetic approaches. Genetic diversity among platyopuntias may enable the low-temperature tolerance of *Opuntia fragilis* and *O. humifusa*, which are native to Canada, to be bred into species from Mexico with high biomass productivity, such as *Opuntia amyclea, O. ficus-indica*, and *O. streptacantha*. Biotechnology can also help improve the disease and insect resistance of agaves and cacti.

In addition to results from breeding and selection, other economic advances can be based on existing cultivated plants. The agave-fiber industry can be revitalized when the secondary products of leaf pulp and sapogenins are fully commercialized. Cactus pears and nopalitos should greatly increase in popularity, and new cactus fruits will undoubtedly be introduced into world markets. Cattle will increasingly be fed agave leaves and cactus stems specifically raised for forage and fodder. Besides their present uses as hedges and fences (Figures 1.7C and 3.7), agaves and cacti may be planted to stabilize land and to prevent erosion in arid and semiarid regions. The loss of top soil is a growing problem, and so plants like agaves and cacti that can survive on rocky, less fertile soil and steep slopes can be important land stabilizers in agriculturally marginal areas.

Although less dramatic to most people, our physiological understanding of agaves and cacti will advance. Techniques should be perfected to enable the visualization of root growth and development in the soil. Interactions between stresses, such as high light and high temperature, will be understood at a cellular level. The induction of new proteins in response to water stress (Figure 5.1) or to low temperature (Figure 5.4) should be understood at a molecular level, paving the way for biotechnology and breeding efforts. Further knowledge about the role of hormones in agaves and cacti, such as for the development of new roots or flowers, will immediately have many agricultural applications. Micropropagation techniques for agaves and cacti (Figure 8.1) could become commonplace. Their relatively unstudied vascular system for the transport of sugars, the phloem, should be carefully described, with the possibility of many new surprises about these CAM plants. The currently known responses of net CO_2 uptake to water, temperature, and light for agaves and cacti will be reinterpreted for the newly discovered environmental variations accompanying global climate change.

The recorded story of our association with agaves and cacti begins with the remnants of these plants in 9000-year-old human feces. This archaeological discovery is a rather humble prelude to discussing agave juices that intoxicated priests (Figures 1.1 and 2.4, Plates D and E), the use of prickly pear cacti in the production of cochineal dyes with a brilliant red hue (Figure 3.5, Plate I), the burning of cactus spines to enable cattle to browse on the stems (Plate J), and bandits stealing agave cuticles to wrap food for a Christmas delicacy (Figure 2.2, Plate C). We have considered the retention of water by an agave leaf or a cactus stem during a long hot desert summer, the survival of cacti under snow in the winter, and the greater annual productivity of *Agave mapisaga* and *Opuntia ficus-indica* than of conventional crops like alfalfa, corn, rice, and wheat. Whatever you have learned and enjoyed along the way, I hope that you now have a much greater appreciation and understanding of these remarkable agaves and cacti!

Index

Page numbers in **boldface** indicate where the entry is defined or illustrated.

Abscission, 75–76, 81
Acclimation, temperature, 102–104, 131
Active transport, **86**, 118, 143
Adenosine triphosphate (ATP), 114, 115, 118–119
Agave, 5, 7, 20, 136, 137. *See also individual Agave species*
  leaf angle, 20, 22, 94–95, 157
  roots, 74–75, 80, 82, 84–85
  uses, 28–44, 128, 147
Agave americana, 5, **6**, 41, 42
Agave angustifolia, 12, 35, 39, 42
Agave atrovirens, 30, 41
Agave attenuata, 23, 42
Agave cantala, 40, 42
Agave deserti, **13**
  CO_2 uptake, 122–125, 129–134
  flowering, 14, 143–144
  growth, 87, 135–137, 139, 143–144
  roasting pit, 28, **29**
  roots, 74–75, 79, 84
  seedling establishment, 14, 104–106, **105**
  temperature, 100, 102
  transpiration, 122–125
Agave filifera, 42
Agave fourcroydes, 38–41, **39**, 43, 138, 139, 151
Agave lechuguilla, 38–39, 42, 44, 84
Agave lurida, 40, 42
Agave mapisaga, 75
  beverages, 30, 32, **33**, 41
  CO_2 uptake, 109, 121–123, 125
  productivity, 75, 140, 151, 159

Agave's, other, 11, 35, 39, 42
Agave parryi, 100, 102
Agave salmiana, 41, 42, 43
  beverages, **4**, 12, 30, **31**, 32, 35
  productivity, 140, 151
Agave sisalana, 4, **31**, 38, 40–41, 138, 139, 151
Agave tequilana, 30, **36**, 37–38
Agave utahensis, 100, 102
Agave victoriae-reginae, 10, 42
Agave vilmoriniana, 11, 42–44, 151
Aguamiel, **32**–34, 152
Air gap, root-soil, 80–82
Alcoholic beverages, 19, 28. *See also* Mescal; Pulque; Tequila; Wine
Alkaloid, **60–61**, 65
Anatomy, **3**
Apical meristem, 20, **22**, 36–37, 89, 95, 97, 98, 150
Apoplastic pathway, **76**, 80
Areole, **20**, 53, 149
Argentina, 7, 48, 49
Arid, **5**, 73, 94, 103, 112, 123, 128, 143, 147, 156, 158
Ariocarpus, 61, 63
Ariocarpus fissuratus, 104, 141
Arizona, 8, 15, 20, 28, 53, 96, 99, 105
ATP, 114, 115, 118–119
Australia, 4, 38, 46, 56, 65–67, 148
Aztec, 3, 4, 32, 46, 58, 60

Barrel cactus, 17, 20, 22, 45, 55, 62, 84, 90, 95, 98–100, 122. See also

Barrel cactus (*continued*)
  Copiapoa cinerea; Ferocactus
  tilting, 14–16, 95, 97
Biochemistry, 108, 110, 115, 120
Biogas, 62, 154
Biomass production. *See* Production;
  Productivity
Biznaga, 45. *See also* Barrel cactus
Bolivia, 48, 49
Boron, 85, 86, 135
Brazil, 40, 49, 58
Bulbil, **12**, 40, 41, 42, 149
Bundle sheath cell, 112–113, 118
Burbank, Luther, 56, 66, 67

C_3 plants, 110, 117, 118–120, 141
  CO_2 uptake, 121–123, 125–126, 132,
    156
  productivity, 141–142
C_4 plants, 111, 118–120
  CO_2 uptake, 121–123, 125–126, 132,
    156
  productivity, 141–142
Cabeza, **30**, 35–37, **36**
Cactus, 5–7, 20. *See also individual cactus*
  species
  leaves, 62
  roots, 74–76, 82, 84
  uses, Seri Indians, 17–19
Cactoblastis cactorum, 67–68
Cactus pear, **46**–51, 56, 158
Calcium, 57, 85, 86
California, 10, 25, 29, 48, 55, 56, 59, 64,
  68–70, 97, 147
Calvin (Calvin–Benson) cycle, 110–111,
  112
CAM. *See* Crassulacean acid metabolism
Canada, 100, 147, 158
Canary Islands, 42, 58, 59, 60
Candy, cactus, 45, 63
Carbohydrate, 30, 83, 84
Carbon dioxide, 24, 82, 109–110
  atmospheric level, 62, 117, 156
  fixation, 108, 112, 114, 117–120, 131,
    138
  uptake (exchange), 85, 90, 108–109,
    120–126, 129–137
Cardón cactus, 17–19
Caribbean islands, 38, 39
Carminic acid, 59
Carnegiea gigantea, 7, 12, 20, **21**, 65
  morphology, 95, 98–99
  roots, 74, 75
  seedling establishment, 104–105

temperature, 98–99, 101
  uses, 17–19, 45, 47, 51, 61
Cattle, 55–58, 151, 154–155, 158
Cell membrane, 76, 93
Cell wall, **76**–77
Cellulose, 76
Century plant, 10, 14. *See also individual*
  agave species
Cereus, 52, 87, 149
Chihuahuan Desert, 86, 104, 135, 141
Chile, 7, 15, 25, 47, 48, 49, 59, 97, 98,
  99, 105, 128, 140, 141, 148, 154,
  155
Chlorenchyma, **24**, 94, 100, 102, 104,
  112–113
Chlorophyll, 24, 94, 100, 134
Chloroplasts, **24**, 94, 104, 112–113, 114,
  115, 116, 118, 143
Cholla, 17–18
CITES (Convention on International
  Trade in Endangered Species), 10,
  64, 148
Cladode, **12**, 53–58, 100, 122, 144, 152,
  156
  orientation, 95–97, 133
CO_2. *See* Carbon dioxide
Cochineal insect, 47, 58–60, 66–68, 69–
  70
  dye, 58–60, 159
Colombia, 52, 153
Colonche, 48, 51
Columbus, Christopher, 47
Columnar cacti, 55, 65, 95, 98
Compass plant, 15
Competition, 106, 137, 147, 148
Convergent evolution, **7**
Copiapoa cinerea, **15**, 75, 93, 97
Cortex, root, **76**, 80, 86
Cortisone, 42–44, **43**, 151
Coryphantha, 61, 64, 104, 141, 150
Coryphantha vivipara, 9, 100, 101, 102
Crassulacean acid metabolism, 26-**27**, 83,
  108
  CO_2 uptake, 121–122, 125–126, 132
  pathway, 111–113, 118–120
  productivity, 120, 139–142, 147
Cross-pollination, **11**
Cuticle, **24**, 30–32, 90, 91, 92, 93, 100
Cylindropuntia, **7**
Cytosol, **112**–113

Dactylopius, 58–**59**, 66–68, 69
Decarboxylation, **112**–114, 118, 120
Decorticator, agave, 39, 41

Diabetes, 34, 46, 64, 153
Dicotyledon, **5**, 23, 24, 75, 92
Diet
  animal, 55–58, 70, 154–155, 158
  human, 3, 16–19, 28, 53, 58, 152
Diffusion, **24**–26, 86, 91, 110, 113, 122–
  123
Drosophila, 65
Drought, 22, 24, 84, 90–91, 93–94, 102,
  105, 126, 134, 143
  and soil water, **80**–82, 129–130, 135
Dry weight, **43**, 83, 128, 138, 141, 143,
  144, 152
Dwarf cacti, 104, 141

Eastern Africa, 40–41, 151
Echinocactus, 45, 63, 104
Echinocereus, 7, 51, 61, 62, 63, **64**
Echinocereus engelmannii, 17, 74, 79
Ecuador, 61
Elevation and agave growth, 135–137
Embolism, 80
Endangered species, **8**–10, 148–151
Endodermis, **76–77**, 80, 85
Energetics, photosynthetic, 114–120, 141
Energy currency, 114–116, 118–119, 143
Environmental productivity index (EPI),
  134–137, 148
Enzyme, **110**, 112, 116
EPI, 134–137, 148
Epidermis
  root, 76, 80
  shoot, **24**, 25, 90, 92, 93
Epiphyllum, 63, 149
Epiphyte, 75, 112, 131, 141
Epithelantha, 61, 104, 141
Eriosyce ceratistes, 99–100
Ethnobotany, 3, 16, 46
Eulychnia acida, 61–**62**
Exotherm, 101–102

Fairy ring, 14
Feces, mummified, 28, 158
Fence, 41, 42, 61–62, 158
Ferocactus, 8, 51, 55, 62
  tilting, 14–16
Ferocactus acanthodes, 20, **21**
  apex, 89, 99
  CO_2 uptake, 122–125, 129–134
  rib number, 22
  roots, 74–75, 79, 80, 84
  shoot, 91
  temperature, 99, 102
  transpiration, 122–125

Ferocactus covillei, 17–19, 47, 99
Ferocactus histrix, 45, 52, 63, 104
Ferocactus viridescens, 98
Ferocactus wislizenii, 17–19, 47, 63, 74,
  97, 99
Fertilizer, 85, 135, 154
Fiber
  agave, 38–41, 151, 158
  synthetic, 40–41, 151
Fibonacci number, 20, 22, 94
Flower, 10–12, 16, 50, 60, 97, 143–144,
  149, 157
  as food, 28, 30, 51–52, 62
Fodder, **55**–56, 58, 67, 154–155
Forage, **55**, 57–58, 66, 70, 154
Freezing, 89, 101–103, 105, 154
Fresh weight, **43**, 57, 141, 145
Fruit, cactus, 16, 20, 45–52, 87, 143,
  152–153. *See also individual species*
  production, 48, 49, 56, 144
Fruit fly, 65
Fungi, 83, 84–85

Gas exchange, 24–26, 108, 120–126
Germplasm, **148**, 151
Glochid, **20**, 46, 53–54, 152
Graft, 13, 149
Growth, 97, 105–106, 108, 120, 128–
  129, 132–133, 135–137, 143–144
Gymnocalycium, 7, 63

Hair. *See* Pubescence; Root
Hallucinogens, 60–61
Hardening, temperature, 102–104
Hecogenin, **43**–44
Hedge, 41, 42, 48, 62, 65, 67, 158
Henequen. See *Agave fourcroydes*
Heyne, Benjamin, 26–27, 108
Hormone, 65, 145, 158. *See also* Steroid
Hydraulic conductivity, 78, 81. *See also*
  Root; Soil
Hylocereus, 51, **52**, 153

Ice crystals, 102–103
India, 62
Indian, 16–19, 28, 29, 45, 47, 60, 61
Inflorescence, agave, **12**, 143–144
  as food, 16, 19, 28, 41
Irrigation, 51, 128, 140, 143, 153
Israel, 20, 47, 48, 49, 52, 96, 97, 153
Italy, 25, 47, 49–51, 148

Jalisco, 36, 37, 51
Joule, **115**

Kenya, 4, 8, 147

Leaf area index, **137**–139
Lechuguilla. See *Agave lechuguilla*
Light, 16, 94–97, 132–134
Lophocereus, 95, 17–19, 65
Lophophora williamsii, 60–61, 75
Lysergic acid diethylamide (LSD), 61

Macronutrient, 85-**86**
Madagascar, 6, 7, 8
Magnesium, 85, 86
Maguey, **29**
Malate, 111, 112–113, 114, 118, 133
Mammillaria, 7, 17, 18, 51, 61, 63, **64**,
    104, 141, 153
Margarita, **37**–38
Medicine, 18, 64–65, 152–154
Mediterranean, 47, 51, 58
Megapascal, 77
Melcocha, 48, 51
Membrane, cell, 76, 93, 102, 103
Meristem. *See* Apical meristem
Mescal, **28**, 30, 34–37, 152
Mescaline, 60-**61**
Mexico, 3, 4, 6, 7, 12, 13, 16, 28, 30–41,
    46–49, 51, 53–55, 57–60, 64, 109,
    140, 141, 153, 158
Micronutrient, 85–**86**
Miel de tuna, 48, 51, 153
Mitochondria, 114, 115, 116, 143
Mixiote, **30**, 31, 53, 147, 153, 159
Mole, **115**, 132
Monocarpy, **14**, 144
Monocotyledon, **5**, 22, 23, 75, 92
Morphology, **3**, 94–100, 103, 121, 157–
    158
Mucilage, 64, 103, 153–154
Mycorrhizae, 84–85
Myrtillocactus, 51, 62

NADP, 114, 116, 118–119
Native Americans, 28, 45, 46, 60, 147
Neobuxbaumia, 55, 104
Neoporteria, 7, 75
Nicotinamide adenine dinucleotide
    phosphate (NADP), 114, 116, 118–
    119
Nitrogen, 57, 85, 86, 106, 135, 154
Nopalea, 53
Nopales, 3, 53, 55
Nopalito, **30**, 45, 53–55, 85–86, 140,
    147, 153
Nurse plant, 104–106

Nutrient, 63, 73, 82–83, 84–86, 89, 106,
    135, 154
Nutrition, animal, 57–58, 83, 86, 151,
    154–155, 158

Oaxaca, 35, 51, 59
Opuntia, 20, 149. *See also individual
    opuntia species*
Opuntia acanthocarpa, 80
Opuntia amyclea, 47, 139, 149, 158
Opuntia aurantiaca, 67, 68
Opuntia bigelovii, 7, 17–18, 55, 95
Opuntia cochenillifera, 58
Opuntia cylindrica, 61
Opuntia engelmannii, 17, 55, **56, 70**
Opuntia ficus-indica, 7, 12, 20, **21**, 47, 62,
    66, 67–68, 128, 149, 157
    cladode orientation, 97, 133
    cladodes as food or fodder, 53–58, 68,
        85–86, 154–155
    CO_2 uptake, 121–123, 125, 129–134
    cochineal, 58–**59**
    fruit, 20, 47, 49–51, 68, 144–145,
        152–153
    growth, 87, 97
    productivity, 138, 139–141, 154, 158–
        159
    roots, 76, 80, 84
    stomates, **23**, 92
    temperature, 101, 103, 104, 147
Opuntia fragilis, 100, 102, 147, 158
Opuntia fulgida, 17, 18, 48, 55, 62, 153
Opuntia humifusa, 100, 102, 103, 132,
    158
Opuntia joconostle, 47, 48, 64
Opuntia leptocaulis, 61, 104
Opuntia littoralis, 68
Opuntia megacantha, 47, 70
Opuntia oricola, 68, **69**
Opuntia's, other, 45, 47, 53, 66, 74, 103
Opuntia phaeacantha, 17, 55, 62, 70, 96–
    97
Opuntia quimilo, 87
Opuntia robusta, 48, 53, 149
Opuntia streptacantha, 47, 48, 53, 149,
    158
Opuntia stricta, 57, 62, 64, 65, 148, 154
Opuntia tomentosa, 58
Opuntia versicolor, 17, 45, 55
Opuntia violacea, 17, 19
Organ pipe cactus, 17–19
Orientation, shoot, 94–97
Oroya peruvianus, 100
Oxygen, 82, 114, 116, 117, 119

Pachycereus, 61, 64
Pachycereus pecten-aboriginum, 47, 51
Pachycereus pringlei, 17–19, 47, 51, 55
Parenchyma, water storage, 24
Parodia, 7, 11
Pathway, photosynthesis, 110–123, 126, 156
Pear burner, 57, 70
PEPCase, 110, 111, 112, 118, 124
Pereskia, 7, 112
Perfume, 62
Peru, 59, 60, 61
Peyote, 60–61, 75
Phloem, **143**–145, 158
Phosphoenol pyruvate carboxylase, 110, 111, 112, 118, 123, 124
Phosphorus, 57–58, 84–86, 135, 154
Photon, 94
Photorespiration, **114**, 116–120, 122, 156
Photosynthesis, 24, 89, 94–95, 97, 110–112, 116–117
Photosynthetic carbon reduction cycle, 111, 112
Photosynthetic photon flux (PPF), **94**–97, 106, 128, 140
  and CO_2 uptake, 132–134, 137
Phototropism, 97
Physiology, **3**, 97, 108, 147, 158
Pitaya, 51–52, 147, 152, 153
Plasmodesma, **76**
Platyopuntia, **7**, 16, 55, 57, 62, 65, 95–97, 102–103, 120, 134, 137, 140, 148, 154, 156. *See also individual opuntia species*
Poison, and agaves, 42
Pollen or pollination, 10–12, **11**, 149
Polysaccharide, 30, 103, 119
Potassium, 85, 86, 135
PPF, **94**–97, 106, 128, 140
Pressure units, 77
Prickly pear, 4, 10, 12, 20, 46, 47, 57–58, 156. *See also individual opuntia species*
Production
  fruit, 48, 49, 56
  nopales, 55
Productivity, biomass, 4, 32, 48–49, 84, 97, 120, 124, 128–129, 137–138, 154, 158–159
  magnitudes, 138–143
Propagation, 12–14, 158
Protein, 57, 110, 158
Pubescence, 20, 95, 98–99
Pulque, 4, **32**–34, **35**, 65, 109, 147, 152

Queso de tuna, 48, 51, 153

Radiation
  light, 94–97
  plants, **6**
Rainfall, 5, 73, 87, 93, 96, 97, 136, 140, 156
  and soil water, 75, 77–79, 129
Ramet, **13**, 14, 149
Rectification, 73, 82
Relative humidity, 26
Religion, 18, 32, 45, 60
Respiration, 114, **143**
Rhipsalis, 7, 8, **9**
Rhizome, **13**, 14, 40
Rib, cactus, 22, 95
Ribulose-1, 5-bisphosphate carboxylase/oxygenase, 110, 112–114, 116–120, 122–123, 124, 156
Roasting pit, 28, **29**, 30
Rocks, and soil water, 79
Root, 22, 128
  anatomy, **76**–77, 80, 84
  distribution, 22, 73, 74–75, 79, 106, 143
  hydraulic conductivity, 78, 80–82, 85
  morphology, 22, 24, 73–76, 79, 81–83, 86–87
  water potential, 80
Root-shoot ratio, 83–84, 128, 143
Root-soil air gap, 80–82
Rubisco, 110, 112–114, 116–120, 122–123, 124, 156

Saguaro. See *Carnegiea gigantea*
Salinity, 83, 86–87, 135, 154
Santa Gertrudis cattle, **56**
Sapogenin, 40, 41, 42–44, 151, 158
Scozzolatura, **50**–51, 153
Seawater, 86, 87
Seed, 10, 12, 13, 19, 65, 73, 143, 152
  cactus, as food, 17, 45, 46–48, 153
  germination, 73, 80, 105
Seedling, 14, 104
  and nurse plants, 104–106
  water relations, 22, 91, 93
Semiarid, **5**, 73, 103, 112, 123, 128, 143, 147, 156, 158
Senita. See *Stenocereus thurberi*
Seri Indians, 16–19, 63
Serotonin, **61**
Shoot, 89–91
Sina. See *Stenocereus alamosensis*
Sisal. See *Agave sisalana*

Smoke signaling, 18, 29
Societies, 19–20, 64, 148
Sodium, 57–58, 85, 86–87, 135
Soil, 41, 51
    hydraulic conductivity, 78–**79**, 81–82
    nutrients, 85, 86, 135, 140
    temperature, 74–75, 104, 105–106
    water, 77–82, 105, 126, 128, 141
Sonora, 35, 57, 99
Sonoran Desert, 7, 13, 45, 47, 74, 86,
    122–123, 131, 135, 139, 153
South Africa, 7, 47, 48, 49, 56, 58, 59,
    67–68, 148
Sphere, area and volume, 22, 91
Spineless cacti. *See* Burbank, Luther;
    Opuntia ficus-indica
Spines, 45–46, 65, 89, 95, 98–99, 123,
    141, 148, 149, 154
Starch, 30, 37
Stem area index, **137**–139, 140, 154
Stenocereus, 47, 51, 152, 153
Stenocereus alamosensis, 17, 65
Stenocereus thurberi, 17–19, 47, 65
Steroid, 42, 44, 65
Stomate, **23**, **24**–25, 91, 92, 121, 123
    opening, 26, 90, 108, 113–114, 122,
    124, 126, 129, 133, 134
Suberin, **77**, 80, 82
Sugar, 30, 37, 48, 57, 63, 84, 103, 110,
    143–144, 154
Sun, 96, 133, 134
Supercooling, **101**
Symplastic pathway, 76–77, 80, 86

Tamaulipas, 38, 39, 51
Tanzania, 4, 40, 41, 147
Taproot, **73**–74, 75, 84
Taxa, **5**
Teddy bear cholla, 17–18, 95
Temperature, 16, 97–104
    and CO$_2$ uptake, 130–132, 134, 136–
    137
    low, 89, 98, 100–103, 132
    high, 89, 98, 103–104, 132, 137
    shoot, 89, 90, 94, 124
    soil, 74–75
    and water vapor, 25–26

Tephrocactus floccosus, 100, 132
Tequila, 34–38, 152
Texas, 55, 56–57, 60, 70, 154
Threatened species, **8**, 148–151
Tissue culture, **149**–150
Tolerance
    drought, 93–94
    high temperature, 103–104, 105
    low temperature, 100–103, 105, 132
Transpiration, **26**, 89, 90, 92, 97, 100,
    123–126
Tribe, cactus, **7**, 51
Trichocereus, 7, 61, 100, 105
Tunisia, 49, 58

United States, 7, 13, 16, 28, 30, 46, 48,
    49, 51, 53, 55, 148, 156

Vacuole, **112**–113
Valley of Mexico, 30, 109
Vascular tissue, **29**. *See also* Phloem;
    Xylem
Volume : area ratio, 22, 89, 90–91, 93,
    129–130

Water, 14, 63, 73, 115–116, 141
    and CO$_2$ uptake, 129–130, 134, 136
    conservation, 24–27, 73, 83, 89, 92,
    114, 126
    energy. *See* Water potential
    freezing, 101–102, 103, 142–143
    in shoots, 57, 62, 90
    storage, 75, 90–91, 147
Water potential, **77**–80
    plant, 80, 90
    soil, 77–82
Water relations, **89**, 92
Water-storage parenchyma, **24**, 90, **93**–94
Water-use efficiency, 108, 124–126, 147,
    156
Water vapor, 25–26, 124
Wine, 19, 48, 51

Xylem, **76**–**77**, 86, 90

Yucatán, 39, 40